WILDLIFE
ON YOUR DOORSTEP

Relax in your garden and enjoy flocks of screaming Swifts.

WILDLIFE
ON YOUR DOORSTEP

MARK WARD

*This book is for Gracie
who loved the autumn leaves.*

First published in 2017 by Reed New Holland Publishers Pty Ltd
London • Sydney • Auckland

The Chandlery, 50 Westminster Bridge Road, London SE1 7QY, UK
1/66 Gibbes Street, Chatswood, NSW 2067, Australia
5/39 Woodside Avenue, Northcote, Auckland 0627, New Zealand

www.newhollandpublishers.com
Copyright © 2017 Reed New Holland Publishers Pty Ltd
Copyright © 2017 in text: Mark Ward
Copyright © 2017 in images: Mark Ward
and other contributors as credited on page 173

A record of this book is held at the
British Library and the National Library of Australia.

ISBN 978 1 92151 774 7

Group Managing Director: Fiona Schultz
Publisher and Project Editor: Simon Papps
Designers: Diana Russell and Andrew Davies
Production Director: James Mills-Hicks
Printer: Toppan Leefung Printing Limited

10 9 8 7 6 5 4 3 2 1

Keep up with New Holland Publishers on Facebook
www.facebook.com/NewHollandPublishers

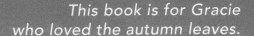

Few sights in the British
countryside are as glorious as
a field of Common Poppies.

Deer, such as this Roe Deer, are often found in close proximity to human settlements. Dawn and dusk are the best times to see them.

CONTENTS

Search your local Chaffinch
flocks for Bramblings between
October and April.

Foxes will be living on your doorstep –
look out for cubs from April.

INTRODUCTION

The UK is home to a spectacular array of wildlife, but you need only to step outside your back door to enjoy the best wildlife watching of all. I've been lucky to travel to all corners of the UK in search of rare and spectacular creatures, but the species that live on my doorstep – in my garden and in the fields, hedgerows, lakes and woods that surround it – give me the most pleasure by far.

Watching the wildlife that lives alongside me provides a rollercoaster ride of emotions every day of the year. I laugh when the beady-eyed Starlings squabble over the fat balls in my garden and the Badger cubs topple over when they sneeze too hard. I worry if the Blue Tits are bringing enough caterpillars to their chicks in the nestbox and if the Swallows will return safely from Africa. And I am full of envy when somebody sees something in my patch that I haven't!

The more you watch, the more you'll want to see and discover. My obsession with my local wildlife has certainly tested the limits of my wife Laura's patience, so this is an early warning about the effect it could have on your personal life! Blundering round the bedroom in the half-light for a dawn departure to the reservoir to see what migrants have arrived and dropping everything to race off following the news of an exciting arrival are now expected behaviour.

The aim of this book is to inspire you to look at your local wildlife and to view your own local area with new enthusiasm and fresh eyes. Explore new places, look for new things and take advantage of everything the UK's varied seasons bring and you'll find hundreds, even thousands, of new species, becoming an expert on the wildlife of your patch in the process.

Whether you have lived in your home for decades, or recently moved somewhere new, there's a world of wildlife discoveries to make. Watching the same area regularly gives you an unrivalled insight into the secret lives of the creatures that live alongside you. You'll be able to predict the arrival of exciting migrant birds using the weather and see the effect it has on the behaviour of your resident wildlife. Your garden, what you

The field next to my house has more than 50 species of wildflowers.

Me with a Scarce Chaser. Get close to your local wildlife – and enjoy it!

Letting our back lawn grow long turns it into an insect haven.

can see there and the things you can do to attract more wildlife to it is a key focus of the book. Look beyond the garden fence though, and start to explore the wider area and the adventure really starts.

The advice, ideas and tips in *Wildlife on your Doorstep* come from my experiences within a five-mile radius of the chalet bungalow where my wife Laura and I live in the south-west corner of Cambridgeshire. It's a typical patch, with the type of places you can find anywhere in the UK. You don't need to go to nature reserves to see great wildlife – look carefully, and often enough, and you'll find it in the most surprising of places.

Birds, butterflies, mammals and dragonflies feature prominently, but I hope you'll be inspired to discover other families too. Finding, and learning about, the easy-to-overlook 'smaller things' is one of the most rewarding parts of discovering what lives near you.

Instead of seeing a 'bumblebee' you'll get to know up to 10 different species; 'hoverfly' will become dozens of identifiable species and you'll soon be putting names to the many moths that can easily go undetected under cover of darkness. There are dozens of other families though, from snails and slugs to weevils and froghoppers for you to find, identify and keep your patch list growing. With

Our cat, Alfie,
loves watching Badgers

I've also shared seasonal tips and field-craft skills for seeing several sought-after, and tricky, species, which could be found on anyone's patch. Adders, Badgers, owls, canopy-loving hairstreak butterflies and Otters are species I think everyone likes to find more regularly, or see even better. These tips are based on my tried and tested experiences and really do work. In the case of Adders, I went years without seeing one, before finally 'cracking' it one April and stumbling on the right approach to seeing these fabulous creatures.

Wildlife on Your Doorstep shares the secrets of my successes, but it's the inevitable failures that make wildlife watching so rewarding when you do finally strike gold. Seeing my first local Otter after a ridiculous number of hours spent peering across a favourite gravel pit was a huge rush of excitement. I really had to try hard to contain my excitement and avoid scaring it away!

around 70,000 different species in the UK, there is plenty to go at!

I've included a few of the many highlights from my own diaries and some of the photos I have taken on my patch in the first few years of living there. I hope they'll provide entertaining, as well as inspiring, interludes at the relevant times of the year and help set the scene of what it's like to watch the same area regularly.

The seasonal approach to the book will allow you to look for ideas and seek inspiration as and when you might need it, as well as prepping for the seasons to come. Planning is vital, so be prepared for the optimum windows of opportunity for seeing things – and avoid commitments at key times!

You'll be surprised what lives nearby – discovering Badgers on my doorstep has turned them into an obsession of mine.

STARTING OUT

Choosing an area to live, and which house to buy, are among the most important decisions of everyone's life. If you're a wildlife fanatic, the dilemmas over size and proximity to amenities and transport links are complicated by questions such as 'What's the garden like?' and 'What good spots for wildlife are within easy reach?' Wildlife potential was certainly at the forefront of my mind when my wife Laura and I were house hunting.

With dozens of viewings under our belts, we finally found a house we both liked. On that first visit, the then owners made a casual comment that was to have a significant impact on our lives. "Something comes in the garden at night to dig the lawn and eat the vegetables – we think it could be Badgers". Once I'd made polite noises of admiration over the fitted kitchen, I was eager to get outside and investigate this far more interesting revelation! Peering over a straggly privet hedge that marked the eastern boundary of a good-sized plot, I was confronted with a huge arable field, a scrubby bank,

ripe for a Badger sett, and the River Ouse in the distance. My mind was made up: this was the house for us. Fortunately, Laura agreed.

It was a bitter January day when we moved into our new home. As I gazed out across the wildlife-free zone that was our new garden from the kitchen window; over piles of rubble and paving stones and a dying leylandii hedge, I was blissfully unaware this ramshackle spot would soon become a haven for newts, toads, animals, birds and a host of insects – many of which I never even knew existed.

Studying an OS map I'd been given for Christmas one Saturday afternoon got me looking closely at the area we now called home. Footpaths and bridle-ways provided easy access to the river, an ancient wood, fields, a huge reservoir and all sorts of other places I couldn't wait to explore. Looking more closely at the village, I found a churchyard, a pond tucked away behind the playing field and a secluded flood meadow. This put the idea in my mind of setting a patch – an area I could watch regularly and get to know everything that lived there.

Most exciting of all was the chain of gravel pits running north to south right alongside the river. I discovered I could even see these from the garden and was soon precariously balancing on the flat roof extension with a telescope so I could add Coot and other wetland birds to the fast-growing garden list! This was just the start though. I had no idea at that time that this was a major migration flyway and what a range of exciting raptors, terns, waders, wildfowl and gulls I'd be seeing migrating along it from the comfort of my own garden.

I wanted to be able to put a name to everything that lived in this new area and to get to know families I had never looked at in detail before. I set myself a challenge of being able to recognise 75 per cent of the wildlife that could be seen on a one-mile walk from home. You certainly don't have to go to these lengths though, and the best place to start with getting to know your local wildlife is your garden.

Smooth Newts can be attracted to gardens by digging a pond.

MAKE YOUR OWN NATURE RESERVE

Whether you live in a city or the country, or have a small or a tennis court-sized plot, your garden will be home to hundreds of species. Create your very own nature reserve – one where you're the warden – and you'll attract hundreds more.

Watching what is going on in your garden will give you a temperature check on what's going on locally and, with a bit of planning, you can attract in the wildlife that lives in the wider area. Build it and it will come. Even if you have a mature garden, you can keep improving it and adding new features, even targeting

I found Green-winged Orchids growing on my local racecourse.

Learn the 'yaffling' call and you'll see Green Woodpeckers regularly.

QUICK TIP: THE SLOW REVEAL

A lot of shy wildlife appears very early and very late in the day when we have wound down, or are yet to get up, so always open your curtains or blinds slowly and carefully. Badgers, Green Woodpeckers, Muntjac and Roe Deer are all shy creatures that will feed on lawns, but will bolt as soon as they see you. Move slowly when out in the wider countryside too so you don't disturb anything feeding quietly. Peek your head round corners and enter woodland clearings and rides slowly, just in case.

Frogs will reward you with hours of pleasure if you make, and manage, a pond.

certain species that you'd like to have.

Here, you should be able to clock up some wildlife-watching time every day of the year, even if it's just a few minutes staring out the window to see what's visiting the bird feeders. Just doing a round of your flower borders in summer, or standing outside for a few minutes in spring and autumn during migration season (what Laura refers to as surveying my manor) will bring surprising rewards. These are the magic times when anything could drop in, or fly over.

Most people naturally look on their feeders for birds, but remember to look up and listen for those flying over – you'll double your garden list with ease. I've recorded about 40 species perched in our garden, but the overall list is 134, because I also include species that have flown over, or been seen from it.

EXPLORING FURTHER AFIELD

If you have a patch already or haven't yet got one, start with a map as I did and draw on 'your' circle or square with your house in the centre. Make it as big or as small as you like – there are no rules. I chose a circle with a radius of five miles because it included a good mix of habitats, top birding areas and plenty of places I knew I could easily access within 10 minutes of home.

Getting round your patch and into the good bits of habitat is important, so track down access points and footpaths. Regularly visiting or walking the same areas is hugely rewarding, so build them into your walks, runs or even drives to work. All that regular coverage will pay dividends. I do most of my exploration on foot, but a bike is a brilliant way of getting around and seeing and hearing things as you travel. You could even try a canoe if you have a river or canal and fancy getting up close and personal with Kingfishers, Otters, dragonflies, fish and other wetland wildlife.

Look even in the places you might usually walk straight past. Sometimes the most unexpected locations will deliver the goods. You'll find some of our most colourful toadstools growing in lawns, a spectacular array of moths resting around the lights on your house, or even in your local public toilets, and colourful lichens in your local churchyard that look like something from another world.

TOP SPOTS

Every patch is different, but you should be able to track down at least one decent-sized body of water, plus woodland, hedgerows and farmland. Heathland and moorland, not to mention a chunk of

QUICK TIP: COMPASS POINTS

Familiarise yourself with your bearings in your local area to help you find and see migrating birds. Flocks of Swallows, martins and waders piling north in spring will be birds heading elsewhere to breed. Birds flying purposefully south or west in autumn could be immigrants from the continent. You'll know where the sun rises and sets so you can position yourself accordingly, whether you're watching mammals at dusk, birds at dawn or looking for soaring raptors in the heat of the day.

Merveille du Jour – not all moths are brown.

Country lanes are great for wildlife.

coastline, will add more excitement and plenty of additional species. Don't be afraid to explore. You might be granted exclusive access by friendly farmers or landowners who might be grateful for you letting them know what they have on their doorsteps.

Farmland occupies 75 per cent of the land area of the UK, so it's a vital place for wildlife. Farmers and landowners regularly change the way they manage their land. This has a huge effect on how wildlife behaves and what you can see there, such as the arable wildflowers that pop up when the seed bank and soil is disturbed through ploughing.

The field next to my house is ever changing in terms of the amount of land left for wildlife through set-aside and wildflower margins, but also the crops grown in it. Every year provides something new. I'm always peering over the hedge to see what's going on and keeping an eye on what the farmer is doing.

One memorable summer, the wheat crop was red with tens of thousands of Common Poppies growing from the recently disturbed soil. The next, a snowstorm of 1,000 Green-veined White butterflies fluttered low over the crop for a week. In another, there must have been a vole population boom because I saw all five breeding species of UK owl hunting

Explore footpaths through fields and farmland at dusk to track down Little Owls.

Sparrowhawks are a sign of a healthy bird population on your patch – there are three pairs in my village.

over the field that year. None of these sights has ever been repeated.

If you are lucky, you'll have a decent-sized wood in your patch, but even shelter-belts and parkland can be great for woodland wildlife. What's there will depend upon the age of the wood, the management taking place there (for example, coppicing and ride-clearance) and the size, and species, of tree. Even often-scorned coniferous woodlands are worth checking out for specialists such as Crossbill, Coal Tit, Goldcrest and fungi that prefer acid soils.

Hedgerows are sadly far fewer in number now, but are vital wildlife corridors that enable wildlife to move easily and safely between sites, including to and from your garden. A really good hedge is made up of more than five species. It's well worth getting to know your local trees and shrubs and when they bloom, fruit and seed.

Brownfield sites are becoming increasingly important for wildlife, especially invertebrates. Old quarries and industrial sites are among the most important places in the UK for solitary wasps and bees, so don't neglect the 'wasteland' near you. The built environment of towns, cities and villages is also easy to ignore but Peregrine Falcons perch and nest on tall buildings, Black Redstarts may visit on passage, perhaps to nest, and in many areas you are just as likely to see a Fox in an urban street as in a field.

Water really boosts the pool of species available, and the attractiveness of a patch to visiting birds, so find and visit regularly, the biggest bodies of water in your area. If you have lakes or gravel pits, watching regularly will tell you which are best for certain species. Much depends on what food is available. To encourage wading birds to 'stick', you'll need islands, mud and shallow edges for feeding opportunities. At my local reservoir, birds often do not hang around long due to the lack of perching places. It is a huge area of water though and viewed from above, as birds would see it, it is a 'migrant magnet'.

Size isn't everything though. Often elusive mammals – Water Voles and Water Shrews, for example – prefer quieter, smaller backwaters. It is well worth going back to your childhood 'pond-dipping' days to explore the amazing underwater world of pond life.

Tracking down the best areas for wildlife in your patch takes time, but once you've found them, you can concentrate your efforts. Finding a new species is a thrill but stumbling upon a new pond, hedgerow or wood, with its wealth of possibilities, can be just as exciting.

SEASONAL DELIGHTS

Taking full advantage of the varied opportunities provided by the four seasons, and getting to know the times to look for wildlife, will seriously improve your fieldcraft, help you see much, much more and ensure you have targets for every season. Nature doesn't just change guard once a quarter though. Every week of the year – in spring and autumn migration periods every day – provides exciting new opportunities and species to find.

Watching the weather and forecasting what wildlife it will bring your way is a key skill to master. It's great fun trying to predict what might turn up and you'll be surprised how often you get it right. Soon, you'll know the classic conditions in your patch for success in finding migrant birds and even particular species. As well as the temperature, look for extent of cloud cover, wind direction and strength and conditions elsewhere in Europe and around the UK. These will all have an impact on what could be coming your way. Don't complain when there is rain, gales or fog – these can bring the most unusual migrants and vagrant birds to your patch.

It isn't just the behaviour of birds that changes due to the weather. If it's too hot reptiles rapidly warm up and so spend less time basking in the open. If it has been raining, mammals find it easier to dig for and find worms; moths are most abundant on hot, humid nights; and you won't find many butterflies and dragonflies when temperatures are low and there is too much cloud cover.

Taking advantage of the peak flight- and emergence-times for butterflies, moths and dragonflies in your area means you'll see the biggest numbers, but also the freshest and smartest individuals. Knowing the peak flowering times for wildflowers allows you to plan in the time to enjoy the colours at their best, such as Cowslips in April, Bluebells in May, Bee Orchids in June and Purple Loosestrife in July and August. Field guides and websites can provide the peak times for thousands of species, so stock your bookshelf well.

LOCAL SPECIALITIES

Wherever you live in the UK – the downs of southern England, the rugged uplands of northern England, the valleys of Wales, or the glens of Scotland – you'll have some very special species to find and enjoy on your patch.

My location in south-east England means that national scarcities such as Nightingale, Smew, Black Hairstreak, Great Crested Newt and Snake's Head Fritillary are a yearly treat and, through local knowledge, I can find them all with

Look for fish, such as this Roach, when the water is clear in spring and summer.

Keep a pair of binoculars near your back door at all times, so if anything interesting flies over your house or garden, you can run in and get them quickly. Those extra seconds saved can make all the difference for making a positive ID. If something interesting is in the garden or on the feeders, you'll also have them close at hand to make the most of the moment.

ease. My in-laws live in Yorkshire, so Red Grouse, Pied Flycatcher and Dipper are local patch birds for them. My parents live in the Fens so they enjoy thousands of Bewick's and Whooper Swans and one of the highest densities of Barn Owls in the UK.

Living on the Ouse Valley flyway, as we do, means that migration periods in spring and autumn can bring exciting birding. The real rarities have included Baird's Sandpiper from North America at my local gravel pits, Long-tailed Skua from the High Arctic at the reservoir and a Red-rumped Swallow from the

Mediterranean frequenting a local farm, alongside pleasingly regular scarcities such as flocks of Black and Arctic Terns.

Other species will be absent from your patch, some puzzlingly so, but you might be able to do your bit to address this. Hedgehogs are sadly very rare in my area, but it is a staple garden mammal for many. Helping the creatures that are rare in your patch is not only great for the conservation prospects of our threatened wildlife, but you'll get a real sense of satisfaction through attracting something new.

HOW TO: USE THE WIND

• Northerly or westerly gales between late August and November blow seabirds such as Phalaropes, Skuas, Gannets and Kittiwakes onto inland waters.

• A southerly airflow encourages birds migrating north in the spring to set off from Africa and arrive in the UK.

• Easterly winds in spring and autumn 'drift' birds across from the continent. Combined with rain or fog to ground them, look for arrivals of Little Gulls, Black Terns, Redstarts – and rarities.

• You can also use the wind to mask your scent from mammals, by staying downwind of them.

CHANGING TIMES

Many species are on the move and expanding their range due to climate change, especially insects. This brings the promise of new species to look for and a chance to monitor their spread through your area. These include the spectacular Silver-washed Fritillary and Scarlet Tiger; Long-winged and Short-winged Coneheads; plus the exotic-looking Wasp Spider.

A recent, very unexpected addition to my patch has been the green-eyed Norfolk Hawker dragonfly. A small, but strong, colony has formed on a tiny fishing lake that's full of the plant where the females lay their eggs, Water-soldier. I used to have to drive two hours to the Norfolk Broads to see this wonderful dragonfly – now they're on my doorstep every June.

It isn't all bad news for our bird populations. Many species are increasing in numbers and range. Herons are experiencing a population boom as wetlands grow and increase. It wasn't long ago that the first Little Egret nests in my patch cropped up in a heronry next to the golf course. Now their rarer cousin, the Great White Egret, is a year-round sight. Spoonbill, Avocet, Osprey, Red Kite and Glossy Ibis are also becoming more common all the time, even though all were once major rarities.

The Arctic Tern's 50,000-mile round-trip annual migration will bring it through your patch in April and May.

One of the greatest success stories anyone can witness on their patch is the Peregrine Falcon. It is now found in the whole of the UK, even in the middle of cities. Watch your local church spire or tower blocks for birds perched high, surveying the hunting potential, or calling loudly overhead. I'm still waiting for my first local breeding record but I often stare at the church tower, thinking how suitable it looks.

The habitats and places on your patch can also change. Unexpected hot-spots can crop up suddenly, sometimes by accident, so take full advantage of these while you can.

The gravel extraction company drained the largest gravel pit near me a few years before we moved in. This simple act exposed tonnes of lovely wet mud and islands, full of microscopic creatures for wading birds to feast on. For several springs and autumns, it ranked among the best inland wader-watching sites in the UK. It was a red-letter time for myself and other local birders as we

Common Poppy – you can build a local wildflowers list into the hundreds.

Resident rarities on my patch include Black Hairstreak.

LISTS AND DIARIES

Birders are well known for their lists, but in recent years there has been a surge of interest in 'All taxa listing' – keeping a list of all the wildlife species and living things you have seen. I hope you'll be inspired to start your own patch list and keep adding to it. I never thought I'd get such a thrill from identifying a tiny spider with only a Latin name that I found in the shower, or finding that there were three different species of slug in our compost bin.

Adding a new species to your garden or patch list is almost as good as seeing it for the first time. Spotted Flycatcher, Nuthatch and Marsh Tit are widespread birds that all occur within a couple of miles of my home, but all are missing from my garden list. I know that if I finally see them, they will be among the most exciting individuals of each species that I've ever seen! Stock Doves flock on the field next to my house, but one spent a week feeding on seed spilt below our bird feeders and it was a treat to have one on my side of the hedge.

Keep a daily diary, so that you have a record of what you see and can compare dates from year to year, as well as numbers, and note down any interesting behaviour. You'll also have valuable records to contribute to the many local and national wildlife-recording societies. Who knows,

found not just odd singles, but flocks of Wood Sandpipers, Spotted Redshanks, Turnstones and Dunlins – sights I'd only expect to see at the top coastal wader scrapes, let alone five minutes from home.

Silver-washed Fritillary – a spectacular butterfly to seek in woods, peaking in July.

maybe one day you'll be writing a book about your experiences! Seeing the first Swallow, frogspawn, Hawthorn flower or the latest dragonfly of the year will become staples in your calendar. Beating your earliest, latest or highest count is a buzz and if, like me, you have a competitive streak, it adds an edge to your wildlife watching.

NAME GAMES

You don't have to be able to identify everything you find to enjoy your local wildlife, but putting a name to what you see is hugely satisfying. Knowing, or checking later, the flight periods and peak times when looking for certain species provides a big clue as to what you are looking at, especially when it comes to insects. Some wildlife has a very narrow window of availability with 'peaks' as short as a couple of weeks. I still have lots of photographs of things that I haven't yet identified, but it's a great job for those long, dark winter nights when you can't get out, so snap away.

A lifetime will not be enough to discover all the wildlife on your patch, so

The sight of a Peregrine Falcon on the hunt will brighten up any day on your patch.

A dozen species of cricket and grasshopper could live near you, including the Oak Bush-cricket.

get out there and start looking and listing. Be warned though, once you start watching your local wildlife, the housework and all those other important jobs will slip further and further down the 'To do' list. If you're lucky enough to have an understanding family as I do, everything will work out just fine.

To identify a Common Toad, look for brown colour and warty skin.

HOW TO: IDENTIFY THINGS EASILY

Take photographs of anything you can't recognise. You'll be surprised how many things, such as wildflowers and insects, you can retrospectively ID from a good photograph. There are some great websites and forums now where you can even post your shots and let experts identify them for you. You don't even necessarily need a good camera. Using your mobile phone's camera is perfect as you'll always have it on hand ready for that unexpected find. I have identified many new species thanks to my phone!

You can identify many hoverflies on your patch from a photograph – this is *Myathropa florea* on Fennel.

Berries are a vital source of food for many birds, including Redwings, in winter.

WINTER

The short daylight hours of winter mean that the wildlife still active above ground has to spend its time wisely – as do those of us who like to watch it.

It's a time for watching the fascinating survival adaptations of your local birds, marvelling at spectacular flocks and seeking special visitors from northern climes. It's a time to do your bit to help your local wildlife through the hard times and for keeping the feeders filled in your garden.

The lack of cover means that signs of elusive mammals are more obvious, and openings to Badger setts and Otter holts are more visible, so get your tracking hat on and do some detective work for the weeks ahead.

Snow and ice bring challenges for wildlife, but wonderful conditions for watching it. Wrap up, pull on your boots and take a lungful of fresh air. Embrace winter and its wildlife wonders won't let you down.

THE STARS OF WINTER

Birds are the most visible wildlife in winter and there are two main groups to focus on: those seeking feeding opportunities in your garden and the refugees that have fled the Arctic winter. The latter come to take advantage of the UK's comparatively mild winter climate and the abundant food in our waters, hedgerows and seed-rich fields.

Further arrivals take place over the course of the winter, especially if the weather turns cold further east in Europe, but most birds are well settled in to their winter territories. This makes their behaviour and appearances much easier to predict.

INVASIONS AND IRRUPTIONS

No two winters are the same for birds. Wildlife winters are characterised by the invasions or irruptions that certain enigmatic species display. In some years, these species can be easy to find in your area, but it could be another 10 years before the event is repeated. Take advantage while you can!

A classic example is one of the UK's most charismatic and frankly gorgeous birds, the Waxwing. I was fortunate that the time when we moved into our house coincided with a 'Waxwing winter' due to a berry shortage on the continent, so we started with a bang when it came to local birds, even adding them to the garden list in week one.

Diary Notes: 5th January
– Waxwings and Woodcocks

As I brushed my teeth this morning, trilling calls drifted in through the half-open bathroom window. I hurriedly pressed the stop button in case I was just getting overexcited by a malfunctioning toothbrush, but no: it was the sibilant trill of Waxwings! Dripping toothpaste all down the stairs (leading to a later admonishment from Laura), I raced outside, pyjama clad, to find a flock of 40 of these plump beauties sitting in Silver Birches on the green. While watching them from the garden later on as they perched on next door's TV aerial, and proudly pointing them out to the in-laws, the sublime went to the ridiculous as a brace of Woodcock suddenly hurtled through right over our heads, presumably disturbed by the farmer at work in the next field. The distraction of having one of my favourite birds available from various windows in the house is an unbelievable treat, but has done me no favours when it came to getting the boxes unpacked for moving in!

Waxwings appear in sporadic winter invasions – and feast on berries.

It's always worth sussing out where berry-bearing trees and shrubs such as Rowan, Guelder Rose, Cotoneaster, Pyracantha and Hawthorn grow, because you can target your search for Waxwings (together with the much more common Fieldfare and Redwing) if you know that numbers have arrived nationally. Waxwings often frequent the most sombre surroundings, such as car parks, central reservations, roundabouts and industrial estates, because they are often planted up with berry-bearing shrubs. The hedge outside a well-known fast food restaurant held a big flock of more than 80 Waxwings in that first 'Waxwing winter' and our local supermarket has also had them in the car park to liven up our weekly shopping trips.

Short-eared Owl is another species that has an enigmatic presence. In some autumns it arrives across the North Sea in good numbers, leading to a winter of relative abundance, but it can be very scarce in other winters. Where there is one of these golden-eyed beauties, there are often more. They are attracted by a bountiful supply of Short-tailed Voles, which have a cycle of population peaks and troughs. Six 'Shorties' once spent several weeks in an innocuous-looking set-aside field just up the road from my village, but they never returned. The small mammal boom, combined with some wildlife-friendly set-aside left by the farmer, created a one-off event.

Two much more rare and exciting winter birds that could turn up literally anywhere in good winters, and often do, are Great Grey Shrike and Rough-legged Buzzard. A much more frequent bird that fluctuates widely in numbers on a local scale is the Brambling – my favourite bird. If the beech mast crop is poor in continental Europe, but good in your local woods, get down there to look out for their white rumps among the green-rumped Chaffinches as they fly up into the trees.

COLD SNAP TREATS

When temperatures plummet and snow arrives, life gets much tougher for wildlife. These conditions are some of my favourite for getting out and watching what my local birds and animals do. Milder winters are becoming the norm, so when snow is on the ground, set off on foot to explore. There is nothing quite like making the first footprints in virginal snow, knowing you are the only human to pass that way – and finding freshly-made tracks made by other wild creatures.

Icy weather can bring unforgettable close encounters with mammals as they lose their fear in order to exploit new feeding opportunities. If waterbodies freeze over, Foxes patrol the edges looking for sick or weak waterbirds. I always get a buzz every time I see a Fox. There is something particularly special about an encounter in the snow, when that stunning rufous fur-coat glows bright orange against the white world.

Freeze-ups are also a good time to look for Otters, so wrap up and get yourself in a good position overlooking either a local river or lake. Otters are now found on virtually every river in the UK, so they will almost certainly be present not far from you. They're now just as much an urban species as a rural one and you are just as likely to find them in a town centre as on a quiet river in the country. Watching for Otters is a challenge, but the thrill of an encounter with one makes the effort well worthwhile.

When ponds, ditches and lakes ice over, two elusive birds that may spend most of the year undetected in your patch can put on a show. Scan reedy edges for the scuttling Water Rail and the king of camouflage, the Bittern. Water Rails are commoner in winter and are more widespread than many people think. The Bittern is a species on the rise thanks to

Winter is a great time for finding and watching your local Otters.

HOW TO: SEE OTTERS

• Look for fishy spraints on mud and rocks and the scales and bones of half-finished fishy meals. Once you've found an Otter territory, it's just a matter of putting in the time.

• Early and late in the day are prime times. Concentrate on the quieter backwaters.

• Rivers are classic habitat but check lakes, gravel pits and even small pools. Otters will fish in surprisingly small waters if fish populations are good.

• Keep an eye on ducks, swans, geese, and coots. If they suddenly start swimming in the same direction at pace and with purpose, or scattering off the bank there could be an Otter. Scan quickly for that dog-like head sticking up. Gulls can also behave like this in the presence of an Otter.

• Choose calm days – it's easier to see the 'wake' left in the water.

conservation measures and the creation of many new and large reedbeds. I don't yet have them breeding locally, but a 'booming' bird sang for two days from a tiny patch of reeds one March, so I hope it will join our list of breeding birds soon.

Freeze-ups further afield can really shake things up and bring exciting arrivals to you. If the wind turns to the east or north-east, birds flee eastern Europe, heading to the milder climes of the UK, which is bathed in the wet and warm south-westerly airflow and the Gulf Stream. Wildfowl are among the most obvious arrivals. In winter, it pays to keep a close eye on all your local flocks, even your local Mute Swans, especially those in fields, because yellow-and-black-beaked wild swans can occur: the noisy Whooper Swan and goose-sized Bewick's Swan. You might even get lucky with a beautiful 'V'-formation heading gracefully overhead en route to a traditional wintering site.

It's also worth keeping tabs on your local goose flocks because the resident Canada and Greylag Geese could be joined by genuine wild geese. Scan carefully for the smaller, black-belly-barred Greenland and Eurasian White-fronted Geese with white foreheads and the smaller, chunkier forms of Pink-footed and Tundra Bean Geese. Any geese appearing in a strange

Hard winter weather may bring White-fronted Geese.

spot, where you don't usually see your resident geese grazing, are worth checking out too. The Mallard-sized, coastal Dark-bellied Brent Goose can latch onto flocks of its close relative, the Canada Goose, as can Barnacle Geese.

You should notice increases in many of your local duck species as the weather turns cold and birds from further afield come in to swell the numbers. If you have Pochards and Tufted Ducks on deep water, scan carefully for the normally coastal Scaup. You might also notice more drake Pochards than females. This is the case in many parts of the UK because the males stay closer to their breeding grounds while the females winter in southern Europe.

One of my star local winter species, but one that fluctuates dramatically in numbers, is the Smew. This sawbill could turn up on any lake or reservoir in the UK and a 'cracked porcelain', punk-crested, bandit masked drake or a flotilla of tiny, 'red-heads' is a guaranteed winter warmer. Smew are always rare here, but having been to Holland, where I saw a single flock of 1,000 Smew on a small town park

Diary Notes:

11th January – White Nuns arrive

East winds have been drifting across continental Europe for a week now, so I scraped the ice off the car to go and check the gravel pits in case any goodies had been frozen out. Scanning the largest pit close to home, visions of beauty in black and white and red and grey suddenly emerged from the choppy waters. There were five panda-faced adult drakes in total. Looking carefully at the 'red-heads', I could see there were quite a few younger birds in their first winter. One or two Smew have been around all winter at the other end of the complex, but this was a clear influx of new birds.

Wigeon were packed onto the shallow-edged west bank of the pit, mainly sleeping. That was until the air was suddenly filled with the whooshing of wings and the 'wee-oo' of Wigeon whistles. A young Peregrine Falcon, a big bird so a female, made its first of several passes trying to take out an unsuspecting or weaker individual in the flock. No joy today but such a magnificent, athletic specimen as this won't need to wait long I'm sure.

lake, I know that large populations are not too far away. It just takes a freeze-up on the continent to bring them our way. The best time to search for Smew is after Christmas when the hard-weather influxes occur, so search on any areas of deep water close to your home right up until March. This is a species where it's worth keeping an eye on reports around the UK. If you know they're arriving elsewhere, it's the green light to get searching.

Another bird that is becoming more common inland, as the number of deep waterbodies increases, is the stately Great Northern Diver. Most winters see one or two at my local reservoir, and in one unforgettable winter there were five! Cold weather combined with strong winds elsewhere tends to bring in divers and the scarce coastal grebes – Slavonian and Red-necked Grebes – a great reward for inland water-watchers. The views are often a lot better than on a choppy sea as well.

FEEDING FRENZY

A lot of the best winter bird action could take place right in your garden. Once birds discover a constant, easy source of food, they'll keep coming back, so the most important job for you in winter is to feed your birds. I couldn't wait to start plotting and planning how to attract birds

Hard weather on the continent could bring Smew to your local waterbodies.

to my garden. Our first winter was a hard one with persistent overnight frosts, so there was an urgency to get food out and start to experiment. We now have a much tried-and-tested menu on offer throughout the year (more of that later!).

The three basic rules for a five-star bird restaurant are to keep your feeders topped up at all times, put out as wide a variety of foods as possible and provide water. Different heights and types of feeder are important as well, in order to cater for the variety of beaks, diets, feeding preferences and degrees of shyness on offer in the cast! One other important thing is to make sure that you can watch your feeders and birds from a comfy spot in your house.

Great Northern Diver – a winter reservoir treat.

HOW TO FEED YOUR BIRDS

Start with the staple 'feeder', the bird table, and find a good spot for it, not too close to cover where cats can lurk but not too far away so birds don't feel too exposed. Those with roofs keep food dry so it will last longer, and a lip prevents seeds from blowing away. Our main bird table has a built-in flowerbed at its base so any fallen seeds can germinate. It's a good way of getting a few extra sunflower seeds as well. We also have a couple of small hanging birdtables and these prove popular, even with larger birds, such as Magpies and Collared Doves that make them swing violently!

Provide a range of feeders and distribute them widely around your garden. You can buy ones on a pole, but I find that hanging ones encourage birds to stay longer. There are seed feeders with feeding ports, mesh cage peanut feeders (good for Great Spotted Woodpeckers), suet cake feeders and square cages and tubular cages for fat balls. Give them a try and see how you get on. Experimenting will bring the best results.

WHAT TO FEED YOUR BIRDS

The bird food market is now big business, with a tremendous variety to choose from that provides all sorts of other nutritional needs. The garden birds of the

Check all swans in winter – Icelandic Whooper Swans may drop in.

Diary Notes: 21st January   Bittern bonanza

One of my favourite winter days: temperatures below freezing and ice covering all areas of shallow water. With thermals on, I headed to the lagoons to see if the hard weather had forced any reedbed skulkers into the open. I'd never seen a local Bittern and although I hadn't heard of any reports, I thought it was worth a look, figuring that I might at least see a Water Rail scuttling across one of the pools.

I climbed the steps to the hide, opened the flap and couldn't believe my eyes. Standing on the reedbed edge was a Bittern, hunched up like an old man cowering in the cold. I watched it for half an hour. It stayed mainly motionless, but occasionally took a few steps on huge-toed feet, always sticking close to the reeds. If I hadn't known it was there, it would have been easy to overlook as it blended in perfectly, melting away against the brown reeds once it stopped moving.

I returned at dusk in the hope of seeing it make a roost flight into the thicker reeds where it would spend the night, grasping several reeds between each foot as it slept. Four different Bitterns made low flights over the reeds, each to a different roosting spot – amazing. Three Woodcocks also tanked past through the trees and Water Rails called all around. Tremendous viewing, but I was glad to get back to the car and get the heater on.

Starlings love to feed on lawns.

A variety of bird tables will attract more birds.

because they contain a high proportion of wheat, which few birds like, and will probably see you infested with pigeons and not much else. Ask your supplier to let you have a look at a handful and ask what is in it. Ideally you want a good proportion of sunflower hearts and seeds, oatmeal and millet.

Our proximity to farmland means that Woodpigeons are a constant sight in the garden. They spend long periods at the birdtable, but don't get too obsessed with policing what takes your food. More often than not, the squirrels, pigeons and Starlings will outsmart you. They are fabulous in their own way, so just enjoy them.

Suet-based products come in an array of shapes and flavours. Some contain fruit, berries, or even insects for extra protein. Suet cakes, bars and fat balls can be put in special feeders. These are really popular with birds and certainly keep Starlings as regular visitors to our garden where a seed-only menu would struggle. Long-tailed Tits love these too, as do wintering Blackcaps. These Blackcaps are different from the ones that breed on your patch in summer and migrate to Africa in winter. The Blackcaps that winter in UK gardens breed in Germany and there is even talk that the birds from this population are in the process of evolving into a different species.

21st century have a lot more on offer than their predecessors, when a few bacon rinds and some hard bread was good enough and they were lucky to get that!

A good-quality seed mix is a must have for a standing table, smaller hanging tables and hanging feeders. Avoid cheap mixes

I've had mixed success with nyjer seed, but it is worth giving it a go. This very lightweight, tiny black seed is popular with three of our most attractive finches: the Goldfinch, Siskin and Lesser Redpoll. If you are in an area with good numbers of the latter two, up your feeding because they are cracking little birds! You'll need a special nyjer feeder with tiny feeding ports.

Start preparing your winter menu back in the autumn by gathering windblown fruit. We have a couple of apple trees in our garden, but you should be able to forage for apples growing wild, or perhaps even benefit from a neighbour donating some of their windfalls. Keep the apples wrapped in newspaper in a dark corner and they should last into the winter when birds will need them. You can scatter the fruit on your lawn, or impale them on branches. Some folk have luck with Waxwings coming to apples spiked on a branch, but you could certainly make a Blackbird, Fieldfare or Redwing's day.

Suet cake and fat balls are popular with a wide range of bird species

Diary Notes: 7th February –
A stranger in the house
House Sparrow numbers have been building up nicely in the garden, thanks to my lackadaisical approach to pruning. We were eating breakfast this morning when something different caught my eye among the grey-crowned males. Rushing for binoculars confirmed the bright chestnut cap and black cheek-spots of a Tree Sparrow – my first ever in the garden. I insisted Laura took a look, upon which she calmly announced that she'd seen it twice the previous week!

Apples are a great way to attract wintering Fieldfares to your garden.

Have at least one berry-bearing tree or shrub in your garden to tempt in the fruit-eaters, especially the five regular species of thrushes. Rowan and Cotoneaster provide berries that are very, very popular with Blackbirds and they are attractive trees too. Get a Holly bush too for year-round foliage, berries and as one of the two key foodplants in the lifecycle of the Holly Blue butterfly. You can see Holly Blues in your garden in two generations: one from April to June whose females lay their eggs on Holly, the other in July to September who lay eggs on Ivy.

Other natural food sources over the course of the winter are provided by plants such as Teasels (great for Goldfinches) and Sunflowers, so leave the seed-heads standing on these for natural bird feeders.

When I saw flocks of Linnets and Yellowhammers in the field next to my house, I got my hopes up for having them as regular visitors to my garden. Once I'd got out in the field and realised that a large area of set-aside and the weedy paths held plants such as Fat Hen, Good

King Henry and docks in abundance, the penny dropped. There was plenty of natural food already there. Our Privet hedge just wasn't going to compete with the mature Elders flanking the field for a safe place to perch either, so I stopped beating myself up and enjoyed them from the garden fence instead.

You can feed your garden birds throughout the year, but winter is the time to go all out. Natural food is easier to find in summer and autumn when fields and hedgerows are full of seeds and berries, so at these times you can ease off. Early spring can be a particularly vulnerable time for your local birds. Birds need to get into peak condition at this time of year in preparation for the rigours of the breeding season, so keep food available well into spring.

UNDER COVER

The fact that birds were not using our garden at first was disappointing considering our rural location. The previous owners did not feed the birds, but I had a hunch that the arrangement of cover in the garden, and general lack of it, was playing a big part.

To make your garden more enticing to wildlife, think about how it arrives and leaves. Can it get in, and out, easily? Which direction do birds and mammals

Grow different species and heights of trees and shrubs to attract a greater diversity of creatures to your garden.

arrive from? Do you have a tall tree or bush where birds can land and see if the coast is clear to come down? Once you've worked that out, provide stepping stones of cover on the way to help them feel more secure and thus more likely to spend time in your garden.

Think about what to keep, what to remove and what to add. It was great to have a blank canvas in my garden, but keeping, and improving, existing features saved a lot of time that would have been needed while waiting for new things to grow, so don't be too quick to cut down or dig out.

Your garden will be part of wildlife corridors for many species, especially mobile ones like birds, so increase diversity by adding habitats or particular trees and shrubs to the mix that aren't found elsewhere in the network.

HOW TO: FEED YOUR GARDEN BIRDS

If you can provide a varied spread, your garden should become the go to place for your local birds throughout the year. Here's what I feed mine:

• Seed – A quality mix, with little or no wheat, on the bird table and in hanging and pole seed feeders.

• Fat balls – High in energy. Just put whole in a special cage feeder where they'll not be stolen by squirrels and can last a good while. Avoid nets – birds get their feet caught.

• Fat cakes – Square shaped suet products in snug square cage feeders.

• Suet pellets – On the bird table and a hanging cage feeder full of them too.

• Fallen fruit – Apples go out in hard weather as natural fruits and berries dwindle.

• Scraps – Nothing wrong with kitchen scraps, so crumbs and old grated cheese go out too.

Leave natural seeds, such as sunflowers, to attract finches, including Bramblings.

Investing in getting the horrible dying leylandii hedge that dominated the view from the kitchen window taken out in winter (when there was no danger of nesting birds) opened up a big new flower-bed. It's now fully planted up with bulbs and plants and comes alive with insects in spring and summer. It also revealed a pleasant surprise: a straggly Elder growing through that evergreen hedge. At the last moment I asked the tree surgeon to leave it and now it has filled out, it is one of the best trees in the garden, full of berries in autumn and sporting creamy flowers in summer. Adding a perching spot where there were none before encouraged more birds to visit, right outside the kitchen window. One of our nestboxes on the back of the house is right next to it. Great Tits quickly took to the box with a better route in to it and now either they, or the Blue Tits, use it each year, giving us some great viewing while we're doing the cooking or washing up.

We didn't lose out by taking out that leylandii hedge. I could see that our neighbour's garden hosted several large, mature leylandii trees, which would provide the nesting and roosting cover for birds and nice warm microclimates for insects and spiders. Conifer-loving birds including Goldcrest and Coal Tit visit our garden regularly, so it's a win-win.

Don't be too harsh with your pruning. I stopped trimming a mature Pyracantha in the garden in year two and it is now the favoured spot for House Sparrows in my street. If you can let things grow wild, they will be better for wildlife. It always pains me to see people removing mature Ivy from the sides of their house. These are fantastic places for birds to roost. Think 'up' especially if you have a small garden, so that you can make the best use of space. Let a climber creep up the side of the house and don't be afraid to plant a tree or two. There are trees for every size of garden (see boxout).

When planning your garden layout, work out where the wind blows most regularly in your garden. South-westerly is the prevailing wind and we are getting more and more gales originating from Atlantic weather fronts. Place your feeders in sheltered spots and help to create shelter with your planting.

I am not a big fan of fences. This may be because two of our three neighbours have had their fences, that form our

HOW TO: CHOOSE THE BEST TREES & SHRUBS FOR WILDLIFE

• Ivy is a must for a good wildlife garden. In autumn, its late flowers are a hive of activity. Many hoverflies, such as the wasp-mimic *Myathropa florea*, and the honeybee-mimics *Eristalis tenax* and *Eristalis pertinax*, flock to it.

• Holly is great too because, along with Ivy, it is the larval food plant required by the Holly Blue butterfly. Year-round foliage, plus berries, make it a must have.

• Rowan gets my vote as a top wildlife tree due to its winter berries, but a good apple or crab apple is also great. Pyracantha, Silver Birch, Hazel, Hawthorn and Cotoneaster will also do the business. Add at least one of these if you have room.

House Sparrows often flock together.

perimeter, blown down by gales and it took a lot of work to replace them. A hedge is easier and cheaper to look after, better for wildlife and looks a whole lot nicer. Air is slowed and can move through hedges rather than pushing against a fence that will inevitably weaken or break. Movement of creatures such as Badgers, Hedgehogs and small mammals is unrestricted too. Beech, Hornbeam, Hazel, Privet, Hawthorn and Blackthorn are all top hedging plants for wildlife.

PEER INTO YOUR NEIGHBOURS' GARDENS

Spend some time seeing what's in the neighbourhood and adjacent to your garden. Once you know what is not far away, you can target those species. Reed Buntings would occasionally fly over our garden, so I made sure that seed was available during the times when the natural seed supplies started to dwindle in the fields (during February and March). They are now a regular late winter and early spring sight, just when the males are getting their smart black heads and bibs.

Have a look at who is feeding their birds, what they're feeding and what birds are nearby. Birds such as Long-tailed Tits will be in flocks all winter and they cover several gardens in their feeding territory. If you know a flock is in the neighbourhood,

Diary Notes:

20th February – Fantastic Mr Fox

I might be biased, but the most handsome dog Fox I have ever seen looks to be a permanent resident in the field next to the house. Every afternoon this week after arriving home from work, I've waited and seen him emerge from the reservoir bank. He takes a good five minutes to cross the open field, before trotting off down the slope for hunting further down the valley. The Rabbits suddenly stop nibbling the grass at the trackside, jump up on their back legs and send the warning signal as he emerges. The Hare doesn't seem bothered, though, and seems to purposefully put on an extra spurt of speed just to show the Fox he has no chance of success.

Listen for Tawny Owls at night, then track down their roosts in the daytime.

they'll pass through your garden at some point, so have a treat ready for them. Suet pellets in hanging feeders are favourites for our Long-tails, as are fat balls. They're great to watch, hanging from them like black-and-white lollipops.

Don't give up because you know that your neighbours are feeding birds and seem to have them all. It's good to have several food sources available and will ultimately attract more birds to your garden.

A NUMBERS GAME

Flocking is a vital survival tactic for many birds in winter and will provide you with exciting viewing to keep out the chill. Safety in numbers means that birds are less vulnerable to predators. In the coldest weather, flocks grow in size as birds arrive from elsewhere in your patch, in the UK or even from the continent, so keep counting.

My best local hard weather flocking experience was when the field next to

Flocks of Long-tailed Tits roam widely in winter, often carrying other species.

the house was covered in snow, but the stubble left from the previous autumn's harvest left plenty of seeds for birds. I saw the biggest flock of Skylarks I have ever seen (more than 500) just 100 yards from my garden: a moving brown carpet of shuffling shapes against the white snow, occasionally taking to the air with buzzing little trills before settling again to ensure they didn't waste too much energy.

Duck numbers shoot up in winter as birds come in from the north and east. An arrival of ducks in winter is a clue to hard weather elsewhere, but also a change in conditions. If Wigeon and Teal arrive on the meadow or gravel pits, I know that the Ouse Washes, a few miles outside my patch, have become fully flooded, making conditions too deep for these surface-feeding 'dabblers' there. Leggy Black-tailed Godwits also come in and pack themselves onto the gravel pit edges and islands when the washes flood. It's worth making counts of your local wildfowl because changes in numbers may reveal some interesting things.

The communal roosting habits of certain birds provide super spectacles

Starling murmurations are a dusk winter treat.

to end a winter's day and even familiar species provide breathtaking sights. Starlings gather (even in town and city centres) in vast smoky flocks that appear like distant smudges on the evening sky. You've probably seen the photographs of such gatherings that the national press love to print, but you almost certainly have a local roost awaiting discovery – and your viewing pleasure.

Crows are easy to ignore, but in winter flocks going to roost can make for a spectacular sight. Keep an eye on where your local birds head off to at dusk. Jackdaws and Rooks may gather in their thousands and it's a sight well worth seeing. The birds make a remarkable sound too.

Pied Wagtails also provide a winter treat. A roost forms in a tree in my busy local market square and they look like Christmas lights when lit up by passing car headlights and shop windows. The most I've seen in this 20-foot high tree is 300 wagtails, so watch where yours go, and which direction they fly as dusk approaches and try and find your local roost. Hospitals and factories are good spots to try.

Another winter-only attraction in my patch is the Goosander. These beautiful 'sawbill' ducks spend the day fishing on the river and are very hard to find and get close to. Each afternoon though, they fly west to the reservoir to roost together. Building up this sort of local knowledge will help you to see the things that casual visitors to your local sites miss out on.

While you are out and about at dusk,

Winter is the time to seek Woodcocks on your patch.

another winter treat is the Woodcock. This chunky, secretive woodland wader arrives in large numbers from Scandinavia in autumn and they melt away in the UK's countryside. Get out at dusk though and you'll find them on your patch. Just standing by an old sludge lagoon near to me, that was surrounded by some tall Ash trees and Hawthorn scrub, produced an impressive count of 21 birds shooting out in all directions over a 10-minute period one evening.

Local patch-watching is about looking in the unexpected areas for wildlife and there are a couple of places that it's easy to turn your nose up at (literally!). Sewage treatment works with the open stirred sewage beds are great spots for Pied and Grey Wagtails to gather, plus Meadow Pipits. These 'microclimates' are warmer than the surrounding area and allow insects to remain active. This is also why overwintering Common Chiffchaffs lurk in the bushes around sewage farms. Watch out for the greyer Siberian Chiffchaff at this time, with its very different 'piping' call. It has been found to be a much more common visitor than once thought thanks to people keeping an eye on their sewage works.

The other grim, but very productive, spot to check in winter is your local rubbish tip. One of the reasons that gulls are much more common inland now is the decline and 'cleaning up' of coastal fishing ports, combined with the increase in

large inland waterbodies for roosting and rubbish tips for feeding. There's no need to enter the tip (which will probably see you ejected anyway). Keep an eye on the surrounding fields where the gulls 'loaf' and digest their unmentionable meals, or scan through the wheeling masses. Any decent gathering of gulls has the potential to hold something unusual and it's a good way of getting to know, and identify, the ages of your local gulls.

My local reservoir holds a huge roost of gulls that peaks in winter with more than 20,000 birds. They fly in from the rubbish tip to the east right over my house, and head out again the following morning. That in itself is a lovely sight to see, but taking a closer look at these birds reveals some hidden winter treasures.

I've seen 14 different species of gulls at my local roost and regular winter treats include those ghostly visitors from the high Arctic – the Glaucous and Iceland Gulls that glow like beacons in the fading afternoon light among their commoner black-backed and Herring cousins. They are well worth braving the chill for.

Even if scanning flocks isn't for you, keep an eye out for the impressively brutish Great Black-backed Gull, which comes inland in winter, and arrivals of the bigger and darker-backed Scandinavian Herring Gulls. Getting to know your paler

HOW TO: SEE WOODCOCKS

There is no better way to end a winter's day than watching Woodcock whizzing over your head in the twilight as they fly out to damp feeding spots to feed.

• Find a spot where you have a 360° view in a clearing, or on the edge of woodland or scrub.

• The point when the first birds spring off the woodland floor is literally the last 5-10 minutes before dark, so don't give up too early!

• Look up against the sky for a chunky, fast-flying form.

• Where there is one Woodcock, there will be more, so keep looking.

• There won't be many other birds flying at this time, so even a poor view should reveal it is a Woodcock.

Search your local gull flocks for scarcities, including the ghostly Glaucous Gull in winter.

local Herring Gulls, which also have more black on the wing, will enable you to pick out these winter visitors from the north. Familiarity with your local Herring Gulls will also enable you to find a scarce visitor from the south, the Yellow-legged Gull, although this species tends to be most frequent in late summer and autumn.

Any flock of birds can be a great place for unusual visitors to lurk in winter, including small land birds such as finches and buntings. Three Lapland Buntings joined a flock of Reed Buntings and Yellowhammers on a recently tilled field on the edge of my village for a week in late winter one year – a real surprise. Look for the larger, paler Scandinavian Mealy Redpolls among Lesser Redpoll flocks. Check out birches and Alders in wet woodland or alongside your local river to find them alongside the wonderfully noisy Siskin and Goldfinch flocks. Their tiny beaks are perfect for extracting seeds from the cones and catkins.

OTHER WINTER WILDLIFE

Winter is certainly the season of birds, but there are other things to see as well. Winter mammal encounters, when many of these animals lose their fear, can be the best of all. Why not take a look at lichens, liverworts and mosses? Get a hand lens to study them to enjoy their intricate beauty. There are also fungi to be found, especially on dead wood, but these are covered in detail in the 'Autumn' chapter, when they really come to the fore.

There are few insects to be found in

Diary Notes:

26th February – My first local Otter

I'd abandoned a trip further afield and decided on a walk along the river instead with nothing in particular in mind to look for, other than to enjoy the walk. As I passed the huge fallen Crack Willow, a spluttering snort came from below me. Looking down, I stared into a whiskery face with a pair of dewy eyes – my first local Otter. It was hard to suppress the heart-pumping excitement and I worried I might spook it by getting over excited. Fortunately, it wasn't worried by me gawking at it and swam along a bit further, scurried up the bank to emerge on the path behind me, giving an unforgettable view as it stared at me again before continuing on to one of the gravel pits for more fishing. Time for some stake-outs on this stretch. The holt must be nearby.

Goldeneyes are a winter visitor to deep water – on warm days look for their 'head-throwing' displays.

mid-winter, although on milder days with a bit of sunshine you might be surprised just after the turn of the year by the appearance of one of the butterflies that hibernates as adults, such as a Comma, Peacock, Red Admiral, Small Tortoiseshell or Brimstone. Leave, or make, a gap in your shed door or walls, and you might find butterflies, and ladybirds, hibernating there in winter.

Check sunny south-facing walls when the hazy winter sun shines and they warm up. It's surprising how many flies can be active, warming themselves up and offering proof that many insects remain on the wing even at this time. This includes several species of moth. My favourite is the fantastically furry December Moth, which you can find resting by lights in walls and porches. More numerous are species such as November Moth, Mottled Umber, Winter Moth and even Spring Usher from January.

Stoats and Weasels are not rare, but they are hard to see. The winter months are among the best times for prolonged

Brush up on your tree ID in winter – find the Alder, find the Siskins and Redpolls.

In warmer winter weather, overwintering butterflies, such as this Comma, may take to the wing.

Squeaking calls reveal the presence of Common Shrews in long grass.

sightings. Most of mine come while I am behind the wheel of my car as they scamper across the road. Always be alert to this and remember the golden rule that Stoats have a black tip to the tail while Weasels are all red-brown. If you see one of these predators scampering across, get your eyes on the tail as quick as you can.

The Weasel is like a high-speed hot dog sausage on legs, low-slung and ferocious. Stoats are larger. Both hunt Rabbits, so it's worth finding, and keeping an eye on, any areas of short turf where Rabbits graze because they could attract the attentions of these miniature predators. One animal that is following hot on the heels of birds expanding their range from the west, the Raven and Buzzard, is the Polecat. Initial reports of them in the east, away from traditional strongholds were greeted with scepticism and dismissed as escaped Ferrets, but they are undoubtedly spreading and could be coming to your patch.

Deer are scarce in my patch and, apart from the odd Muntjac, opportunities for seeing them are limited. If you live in a more wooded area, you could have several species to look out for. Stocky Roe Deer are native and widespread, often feeding in fields, while long-standing introduction the Fallow Deer is very attractive. Red Deer are one of our most magnificent animals and time spent watching

Water Voles are active all year on ponds, rivers and canals.

and staking them out in winter will be well rewarded.

One sound that you you'll hear regularly in overgrown field margins and roadside verges, even in winter, is the squeaking of shrews. It is very high frequency, but with patience you can get a glimpse of these entertaining animals if you follow the squeaking and sit very quietly. With less cover around, winter is a good time to put in some time for a sighting. Pygmy Shrews are our smallest mammal and have to eat more than their own bodyweight each day to survive. You'll sadly find more of these dead than alive, but if you do find a dead one it's well worth having a close look.

You can also set up mini feeding stations in your garden or a local wood for small mammals such as Wood Mouse and Short-tailed (Field) and Bank Voles. These will all come into gardens as well, with the Wood Mouse in particular a regular visitor, so try popping a few seeds or suet pellets in an old log.

Two of our smaller water mammals require patience to find, but remain active in winter as they don't hibernate. Concentrate on quiet stretches of your local river and ponds for Water Voles and Water Shrews. I hadn't realised quite what a fantastic animal the Water Shrew is until one started to appear in late

Stoats are widespread, but you'll need luck and patience to see one.

winter every year in an ornamental pond near me. These are like hyperactive little submarines that plop into the water and swim along in a flurry of tail, limbs and snout under the surface. In clear waters you can follow them but they are creatures of habit so spend some time seeing where they emerge on the surface and be ready with camera, binoculars or just your eyes to see a really fascinating animal.

I use the winter days to get out on foot and seek out wildlife watching opportunities for the excitement of spring ahead. It's a good time for discovering new ponds, lakes or even access points where footpath signs are not hidden by rampant vegetation.

Having ticked off the winter bird specialities in your area, hopefully enjoyed some exciting visitors forced here by hard weather and with a garden full of birds – and nestboxes – for the breeding season ahead, you'll be raring to go for what's to come. By mid-January, you'll notice the light increasing slowly beyond the 4 o'clock dusk of early winter and the countdown to spring begins.

Diary Notes:

28th February – To Russia with love

Although Greylag Geese gather on the fields after the harvest, it's unusual to see geese passing over the house, so when I saw a neatly arranged 'V' of around 35 small grey geese flying high south-west to north-east this morning, I grabbed the bins. They were heading rapidly towards the river, but I just had time to catch the black belly bars of the trailing birds that revealed their identity as White-fronted Geese. Once they hit the river valley, they turned due north to follow it to its eventual termination at The Wash in Norfolk. It's tempting to think they were from the wintering flock at WWT Slimbridge in Gloucestershire, the mild weather telling them it was time to head for their Russian breeding grounds.

Our smallest deer, the introduced Muntjac, has a bone-chilling call.

You'll know spring has arrived when you see your first Wheatear.

SPRING

The day when the first sulphur-yellow Brimstone butterfly
of spring flashes through your garden, or dances along a country
lane, or the deep buzz of a queen bumblebee fills the air confirms
that winter has released its grip and brings the news we've
all been waiting for – spring is here.

This is when the local wildlife year goes into top gear. Every day
brings something different as the temperature rises. The surge of
new life lifts the soul and it's impossible not to be in better spirits
and wanting to be outside as much as possible.

Woods, hedgerows and flowerbeds burst into bloom and a
huge range of migrant birds arrive, looking and sounding at their
best as they set up territory, or just stop off to refuel on their way
further north. Mammals, reptiles and amphibians stir from
their slumber and Mallard ducklings take to the water.

The dawn chorus gets better by the week up until May, so
remember to sit back and relax and enjoy the sounds of spring
as well as the sights.

SIGNS OF SPRING

Spring is the time for looking for local 'firsts'. Note your first Swallow, Blackthorn flower, butterfly, Snowdrop or bumblebee and it will become a yearly tradition. Different people take different things to prove that spring has arrived. When my interest was mainly birds, it was always the first sighting of a Sand Martin feeding over the water, a singing Chiffchaff or a crisply marked cock Wheatear bounding onto a fence post or ridge or furrow in a freshly-ploughed field. The completion of a journey all the way from Africa to arrive on your doorstep is certainly something special to celebrate – and marvel at.

Diary Notes:

9th March – A battalion of Brents

The other local birders often let me know if an interesting bird has flown north through the gravel pits, knowing there's a chance it will continue following the river and be visible from the garden. I got a call this morning to let me know that a single Dark-bellied Brent Goose had just flown north with a flock of Lapwings. Doing a good impression of a headless chicken, I rushed outside with binoculars and fixed them on the valley in the hope of a very rare inland encounter with this saltmarsh-loving goose. Within a minute the tumbling mass of Lapwings were visible but try as I might, I could not see the goose. I was gutted thinking my once in a lifetime chance was gone. Then, the phone went again, a flock of 13 had just gone through! Surely this time? In what seemed like an age, I scanned and scanned above the river, but nothing. Then, there they were 13, chunky charcoal geese powering into the strong northerly wind.

It was a bit of a shock to find a pair of Mallards swimming on our small pond first thing this morning, chatting away to one another. Things could get interesting if they commandeer it for breeding. The male Reed Bunting is looking splendid now with a solid black hood and bib. I give it another week before he goes and switches his seed diet to insects and heads off to breed. I'd love to know if he was a British bird or a continental visitor.

The boundaries between the seasons in the UK are becoming more blurred. We are lucky to have such distinct seasons – this is what brings so many different species to our shores – but spring is coming earlier and milder winters seem to be more common.

Among all the excitement of spring migration, it's easy to concentrate on birds at the expense of all the other riches the season has to offer. There's so much to look for and enjoy, so it's actually quite good that bird migration comes in fits and starts. It doesn't really get going until mid-April, so we'll return to the subject later.

FIRST ON THE WING

The first bumblebees on the wing are the chunky queens looking for places to set up a colony. They emerge from hibernation from late February. Early Bumblebee should in theory be the first, but you could see any of the common species first. There are 27 species of bumblebees on the UK list and it's well worth getting to know them.

For a good bee garden, get your bee hotels up (see the 'Autumn' chapter), plant plenty of nectar-rich flowers, especially in the south-facing beds, and leave some bare patches of soil for ground-nesting solitary bees.

Bees are associated with summer, but there are many to be seen from early spring, including the Tawny and Orange-tailed Mining Bees. One particularly striking species to look for is the Hairy-footed Flower Bee, which is often the first bee on the wing. Watch from March for these charming bees bumbling around your beds. Dead-nettles, Ground-ivy

Male Red-tailed Bumblebees follow the early spring queens.

Brimstone – usually the first butterfly on the wing in spring.

QUICK TIP: RISING BUBBLES

Watch for bubbles suddenly rising to the surface in ponds. Often this will reveal the presence of newts, but it could also help you track down a Water Shrew. Air bubbles are released from their fur as they swim, so follow the bubbles and watch for one breaking the surface. You'll need to be quick though as they travel at quite a lick, both in the water and on land.

The bold black zigzag markings help to identify an Adder – always treat them with respect.

REPTILES AND AMPHIBIANS

and Primrose are good plants to grow to attract them.

Watch out for the well-named Bee-fly in spring. This furry fly has a long proboscis (tongue) that it probes into flowers for nectar. The best place I've found to search for them is in patches of grass near hedgerows where Bugle grows. I copied this and left a big patch in the grass around our greenhouse, which is now frequented by several Bee-flies every March and April. This is a good example of how mimicking habitats further afield can bring new things into your garden.

Spring is the time to search your patch for reptiles and amphibians. I used to mistakenly think that hot days would be best, but when temperatures soar, reptiles soon warm up and head for cover. Concentrate your searches between late February and early May, and put in most effort early in the day.

These are also the months of peak action for our only venomous snake – the Adder. This is one of my favourite species and it is always a thrill to find one curled up, taking in the spring rays. I enjoy hours of safe Adder watching every year and so can you. Just treat them with respect and watch from a distance. Once you've discovered your local Adders, you'll find that they are faithful to the same basking spots, so year after year you can enjoy them.

Once you've tracked them down in your area (see right), the holy grail is to witness the springtime dancing of males. I've only seen this once, but it is a sight that's well worth putting in the time for. Two or more males suddenly head for each other, rear up and slither along at high speed, entwining their bodies.

Spring is an exhilarating time anywhere in your patch, but in the garden it's rejuvenating. As well as enjoying the first sights of emerging insects and the first flowers, it's a time for sowing, planting and adding in new features now the weather is better.

A pond, no matter how small, will seriously increase your garden's potential to attract wildlife. Whenever you are building a feature like this, think about yourself as well as the wildlife you're

Multiple Adders will bask entwined.

HOW TO: SEE ADDERS

• Search on sunny mornings from late February, when Adders bask motionless in the sun to warm up after hibernation. If the temperature is 9°C, it's warm enough to look. Days with some cloud cover and sunny spells are ideal.

• Check south-facing slopes and open areas in heather and other vegetation.

• Once you work out which way is south, you'll soon get into the swing of finding, and covering, likely basking spots. Tread softly and make sure that your shadow doesn't fall across potential basking spots.

• Have your search engine tuned to look for 'dog poo' and you'll be quickly onto any coiled serpents. It really works!

Sweet Violets in the unmown patch on my south-facing front lawn provide early spring nectar.

• Decide on the size based on the size of your garden. Use a butyl liner to shape it how you want with deeper and shallower spots. Or go DIY and put an old sink complete with plug into the ground.

• Make sure that whatever you have has shallow, gently sloping edges. It's better for drinking and bathing birds, but also allows wildlife such as frogs to come and go easily, and means that should a terrestrial animal such as a Hedgehog fall in, it can get itself out.

• Oxygenating plants are a must, so start with those to get the water in good condition. Add colour and structure with plants such as Marsh Marigold, in pots in the water, weighted down with bricks and big stones.

• Water lilies look fabulous, and pond wildlife loves to sit on, shelter under or lay their eggs on the pads (as with snails). Lilies provide the ample surface cover that a good wildlife pond needs.

trying to help. Your reward for helping it is hours of viewing pleasure, so make sure that you can get to it and view it easily.

There were six large goldfish in the pond in our garden when we moved in – a sure sign that this was a pond that had not had wildlife in mind. So not quite a blank canvas, but a great project to get stuck into! I let nature take its course as a Grey Heron, over a period of several months, gradually thinned down the fish population. With the fish gone, other wildlife

Pile logs and stones by your pond as cover for hibernating creatures.

(young newt), confirming they were now breeding.

The more you peer into a pond the more you'll find. Get right down to the water's surface for a really good look and do it regularly. Water boatmen, Great Diving Beetles, stilt-legged pond skaters and Water Scorpions (another garden arrival that brought great excitement!), to name just a few, are tremendous creatures to watch.

was able to colonise without being eaten. Finding the first Smooth Newt one March day resulted in much excited yelling. It got even better when I found the first eft

It's not just birds that are on the move in spring. Common Toad migration is another early spring must-see sight. Toads

Common Toads may colonise your garden pond if you provide cover nearby.

spend a lot of time away from water and hibernate in places such as rabbit holes. Watching them waddle along paths in their dozens in March and April is a wonderful sight. Look for the traditional migration routes on your patch.

Common Frogs also head to their breeding ponds, so peer into your local ponds and keep an eye for the first frog-spawn. Listen for the croaking of frogs from February (rain seems to set them off in particular), building to a spectacular crescendo.

Spring is also a good time to look for newts. Male Smooth Newts are at their best in spring and have impressive markings, colours and dragon-like crests. The much rarer Great Crested Newt takes the honours for most impressive crest, though. Peering into a pond, especially if it is in your garden, is the way to watch newts. Use binoculars to appreciate the fine detail and watch them lithely moving through the water.

SOUNDS OF SPRING

Your local songbirds start tuning up as soon as the days begin to lengthen after Christmas, but once spring arrives the performance goes to a whole new level. The residents start up before the migrants arrive, so put in some quality time to enjoy them as well. There are

Diary Notes:

24th March – Ticking the Velvet

I'm learning that strong, cold winds are the key ingredient for shaking things up birdwise at the reservoir. Following one of the coldest north-easterlies I can remember, I headed over to see what might have been blow in. In my haste, I'd forgotten gloves so after a 15-minute scan from the west end, I was ready to head back to the car, and home to warm up. Suddenly a black duck swam into view, 300 metres away. My eyes were watering and I could hardly turn the focus wheel on my scope but I could see it was a scoter. An excellent find. Suddenly, it dropped its right wing. Was I imagining that flash of white? No, it was a VELVET SCOTER! A county rarity. I'd love to know where this young drake had been born and where it was off to next.

Back home, Black-headed Gulls were piling north up the valley all day long in neat, close-knit flocks. This was no roost movement though. These were birds bound for northern parts, resplendent in their brown-hooded breeding finery.

Smooth Newts look their finest in spring.

Check bare spring fields for Grey Partridges and tune in to their rasping calls.

some beautiful songs to listen for.

I think the Song Thrush sings at its best in the evening, right when other species are winding down for the day. The same goes for my favourite resident songster, the Blackbird. Get outside in the evening in March and April to enjoy them. I can think of no better place to be than sitting in a comfy chair in the garden and letting their notes ease away my troubles – perfect after a hard day at work. You'll be surprised how many Blackbird territories you have in your neighbourhood, so now is a good time of year to map out all the singing, territorial males. The wild song of the Mistle Thrush is just as good. It likes high places from which to sing, even tall radio masts.

Birds are not just making sounds with their vocal organs. The 'drumming' of woodpeckers is at its peak in spring and is a frequent sound in woodlands in March and April. Watch your local Great Spotted Woodpeckers carefully and you'll find that they have favoured spots – usually old hollow trees where the sound of their beaks hammering on the wood at rapid speed will carry furthest, but sometimes on telephone poles, even using the metal plates on them!

Still, sunny days in March and April are the time to listen for the longer, softer 'rattling' drum of the sparrow-sized

The aim of a birdrace is to try and see and hear as many species as possible in a set period of time. May is the month when most species are present on anyone's patch, so it's a great way to test, and improve, your knowledge of your local birds.

• You can do a 24-hour race or a 12-hour, daylight-only one.

• Visit as many different habitats as you can for the greatest variety of species.

• Start early for the dawn chorus and don't forget night time for owls and other calling birds.

• In May, the dawn chorus starts as early as 3am. It really is worth getting up for because the quality and purity of sound is incredible.

• Lots of migrants will be moving in May, so these will boost your list.

• Do some reconnaissance before the big day to pin down key species.

Lesser Spotted Woodpecker, which is a star find on any patch. Listen also for the shrill, falcon-like crescendo from the male accompanying the drumming.

Once April arrives, you'll hear new songs every day as migrants from Africa check in. Why not keep track of how many species you hear in the dawn chorus each morning? It's the perfect accompaniment to a walk to the shops, or a morning jog, and sharpens your birdsong ID skills.

Warbler song will dominate in spring once the migrants arrive. The more habitats you visit, the more species of warbler you'll add to the list. There are wetland warblers, woodland warblers and scrub warblers. Grasshopper Warblers are a star find. Listen for their fishing-reel song early and late in the day, coming from scrubby spots, marshy spots and often wasteground where brambles grow, so check even the most unlikely looking spots. Whitethroats throw themselves off the top of hedgerows and Reed Warblers chunter incessantly from reeds.

ON DISPLAY

Birds rightly win accolades for their song, but their displays are another special springtime experience. Some birds are well known for their displays, such as the 'skydancing' of harriers, and the communal 'lekking' of Black Grouse, but

Great Spotted Woodpeckers reveal their presence with their springtime 'drumming'.

QUICK TIP: NESTHOLE WATCH

Birds can be hard to see in summer once they settle down to breed and the leaves grow, but for hole-nesters such as Nuthatch, Marsh Tit and woodpeckers you can be guaranteed several weeks of viewing if you locate their nestholes early in the spring season. Watch for Nuthatches adjusting the size of their nest hole with mud.

even the common birds on your patch are worth watching. As testosterone levels rise in the males and the urge to defend a territory and find a mate becomes insatiable, things can get pretty spectacular at times.

There is plenty to see in your garden. Greenfinches provide a splash of green and yellow among the House Sparrows at the feeders, but are generally overshadowed by the colours of many of their finch cousins. During spring, though, the males perform a captivating show over parks and gardens. The display starts when males perch on top of a favourite

Look skywards in early spring for displaying raptors, including Sparrowhawks.

Diary Notes: 5th April – Eyes to the skies

Spending time in the garden always increases the chance of an exciting discovery, but today's back-breaking all morning veg patch digging session had pushed my limits of endurance. I stood up to stretch my back and my eyes were instantly drawn to two 'flying barn doors' which circled just to the east. I yelled at Laura, "Don't lose sight of those birds whatever you do" as I ran in to get my binoculars – making a mental note to not have them zipped up in the case at the height of spring migration as they were. I had my suspicions and with bins in hand and Laura directing me to the birds' position using the clouds as markers, I was able to see that they were a pair of magnificent Common Cranes. They circled slowly south down the river valley, using the warming thermals pulsing upwards and enabling them to move without even a flap of their massive wings. Maybe they were wanderers from the not-too-far-away Fenland or Broadland populations, or vagrants from the continent.

We took a break in the afternoon and walked down to the marina. Orange-tips were everywhere and I counted 30 on the walk. Bee-flies were out in force as well. In the corner of one of the pits, Bugle grew in profusion and at least eight of these bizarre insects were hovering at the blue flowers, sticking in their proboscis.

Blackcap in full voice. As the weeks pass in April and May, more migrants join the dawn chorus.

tree, often a conifer. Suddenly they start to spiral up into the air on slow and deliberate wing-beats, which are much slower than normal flight. They twitter loudly and could easily be mistaken for a bat tearing around in pursuit of a fly. Collared Doves perform a simpler, but neat, display, rising from a telephone aerial or tree up at a steep angle, spreading their wings and gliding back down to earth, while the antics of Dunnocks, with their partner-swapping ways, have to be seen to be believed.

Further afield, there are other remarkable displays to look for. March and April are the peak months for raptor display, so watch the weather forecast for sunny days that will tempt them into action. A bit of wind allows the birds to stay in the air without expending as much energy, so they may stay up for longer.

Hopefully, you'll get a Sparrowhawk perched on your garden fence at some point in the year, so you can admire the delicate barring, piercing yellow eyes, needle-thin legs and frighteningly sharp talons. In spring, watch out for pairs rising high above your town, city or village, or over local woods to perform their elastic-winged ascents and headfirst

Diary Notes:

*15th April – It's tough being a mayfly
Laura and I went for a picnic by the river
in town this evening. As we ate, a single
mayfly rose from the long grass between
us and the river and made its first flight
into the air right next to us. I was about
to try and impress Laura by quoting the
fact that mayflies live for only a day when
a single Swallow came pelting towards us,
making me think what a lovely spring scene
it was. It promptly hoovered up the fly with
an audible snap of its beak before jinking
off away across the meadow looking for
its next meal. It was one of those moments
when I wasn't sure whether to laugh or
cry at the brutality of nature.*

*It was a different story along the dam
later in the day with vast clouds of mayflies
making every step a hazardous one – it
was mouth closed time! The emergence
coincided with an arrival of White Wagtails.
Stunning birds which made the poor old
local Pied Wagtails look a little drab in
comparison. There were flashes of sulphur
yellow too among the black and white. Five
male Yellow Wagtails were also feasting
before continuing on their migration.*

plunges back to earth. If you live near remote woods, especially in the north and west, you could see their bigger cousin the Goshawk.

Populations of the Common Buzzard are on the increase and it is now our most common raptor. Choose a sunny spring day with a light wind because it also

HOW TO: SEE GREAT CRESTED GREBES DISPLAY

One of early spring's most delightful attractions is the mating 'weed dance' of the Great Crested Grebe.

• The birds won't perform on tap, so you need to be ready for it and put some time in at lakes and gravel pits watching the birds closely.

• Listen for strange-sounding grunting calls drifting across the water and if you see a pair of grebes swimming towards one another, brace yourself.

• If one has weed in its beak, it will present it to the other one, head-shaking occurs and the full dance will occur with both birds rising out of the water.

The 'weed dance' of Great Crested Grebes is a springtime treat on lakes and gravel pits.

gives you the opportunity to hear their gentle but far-carrying mewing calls. Find an open spot and scan the skies for the flying 'V' silhouette and you might be surprised how many buzzards are breeding on your patch. Scanning across the Ouse Valley from my garden in spring can produce up to five pairs – the only time of year that such numbers are up in the air together. It's a great time for mapping out the territories of your local birds of prey, some of which can be otherwise hard to track down.

Waders perform some of the very best displays of all, especially because they accompany them with beautiful wild and musical songs. Lapwings are birds of marsh, moor, meadow and farmland, and in any of these habitats you have a chance of witnessing their tumbling displays on paddle-shaped wings, flashing black one minute and white the next.

The Curlew has the best song, a liquid bubbling trill that accompanies a parachuting display flight, but the Common Snipe has the most breathtaking display, plummeting down to earth with long beak outstretched and tail feathers spread making a 'drumming' sound like a distant bleating sheep. Check out your local gravel pits and wasteground with puddles for the hyperactive migrant Little Ringed Plover. It too has a great display that sees it winging its way through the sky in a breakneck, spiralling display uttering its grating calls.

Don't forget your winter birds just yet though, as those that linger into spring provide one more treat before they leave. Spring provides a chance to hear their songs and see their displays, from the head-thrown-backwards display of handsome drake Goldeneyes to the chorus of a flock of Scandinavia-bound Redwings in 'subsong' in your local wood.

And it's not just birds that use displays. Butterflies have captivating display flights and pairs of newts take to synchronised swimming in ponds. Mad March Brown Hares career across open fields in early spring. When it comes to the famous boxing, more often than not, it is a female fending off the attentions of an amorous male. While the vegetation is short in early spring, get some quality hare-watching time in. Find the best high spot available so that you can scan several fields at once.

Diary Notes:

26th April – Raising the migration bar

A big north-easterly kicked in overnight, making the conifers sway violently, so I was desperate to be first down to the pits. Surely waders would be on the move? Pulling into an empty car park confirmed that I was first on the scene and no sooner had I set off tramping across the field to the flooded pits than a flock of waders approached from the south. Bar-tailed Godwits. They didn't stop and continued to the far northern edge of the pits, gained height and were off to the north-east. Next stop, The Wash in Norfolk and miles of sandy beaches for them to choose a feeding area.

This was just the beginning. For the next two hours, flocks of this normally coastal wader continued to sweep through. At least two groups stopped for a snooze and a bath, so I could really admire the males in their stunning burnt orange breeding plumage. By the time I left and things had slowed down, the other locals were just arriving and I'd logged 137 birds in total – more than an average year provides in the whole of the county.

Early spring will provide great Brown Hare viewing on your local fields.

WILD ABOUT FLOWERS

There is a tremendous variety in the UK's spring wildflowers, from the show-stopping Snake's Head Fritillary that flowers on an ancient flood meadow in my patch in April to the amazing parasitic Toothwort that grows with Hazel.

One of my favourite places to be in spring is my local wood. The atmosphere in woodland is magical at this time, so make as many visits to your local wood as you can from mid-April to late May, when bird song and the woodland flora is at its best. Search out spots where plenty of sunlight can penetrate the canopy, encouraging flowers to grow. Wood Anemone, the richly-scented Wild Garlic (or Ramsons), Lesser Celandine, Primrose and of course Bluebells provide a fantastic display. The peak flowering time of Bluebells varies depending on how far north or south you are, but April and May are the times to look. If you can track down some ancient woodland, Dog's Mercury and Early Purple Orchid (April to June) might be available.

Flowers attract insects for a large part of the year, even in early spring when only a few species are in flower. The early spring hoverflies were a noticeable gap in my list until I read that Blackthorn blossom was a draw for them when there is an absence of other flowers. Off to the wood

Enjoy the smells as well as the sights and sounds – find Wild Garlic, or Ramsons, in your local woods.

I went on the first sunny day in April, checked the first block of blossoming Blackthorn thickets I found, and I ended up seeing every species in the space of a morning.

Stand close to trees and shrubs bearing blossom to hear one of the best sounds of spring. The effect is incredible with all manner of bees, hoverflies and flies attracted in and creating a buzz. In wetter areas, willows and sallows are just as good, especially in March and April.

Much is made about the colours of autumn leaves, but the green of hedge-rows and woods in May is one of the richest colours of nature and is more than a match for autumn's splendour. A May Hawthorn leaf is one of the most intense greens you could ever see.

Diary Notes:

30th April – A wry smile

News of a Wryneck, that weird neck-twisting, ground-feeding woodpecker, feeding on one of the footpaths by the river at the gravel pits had me hurrying home from work. Surely it wouldn't put up with being disturbed in such well-trodden spot for long? It was still present when I left work, so I hurried round to the river and joined an eager crowd waiting for it to reappear. I spied a movement in the hawthorn by the path and then a small shape dropped down to the short turf below – there it was. It's amazing how wildlife adapts its routine to people. If somebody walked along the path, it just flew up into the hawthorn, waited for about a minute, the resumed its feeding. I later saw that it was feasting on black ants from an ant hill.

As the evening went on, the sweetest sound of the patch started to come from the scrub around the pits. Nightingales are back with 28 territories here this spring. The most intense song comes after dark, but it was good to be able to watch a couple of the males belting out their songs this evening, literally shaking from the effort.

The rare Snake's Head Fritillary growing wild in my patch in April.

As well as searching out flowers in the wider area, it's a good time for deciding which species you'd like in your garden. I grow a good number of native plants that are a characteristic of my local area, including Snake's Head Fritillary, Primrose, Cowslip and Foxglove. Wild-flower mixes are great for borders and lawns and you can, like me, have species in your garden, such as Corncockle, Corn Marigold and Cornflower, which are now very rare in the wild. Don't feel you have to just grow native though: a mix is the perfect solution for a wildlife garden.

There will be great plants for wildlife

Spring bulbs bring colour and nectar in my garden from February to early April.

in any garden. The key is making sure that you retain the best and get rid of the rest. A pleasant surprise in our first March was a dazzling display of spring bulbs coming up in all the flowerbeds – Snowdrops, Grape Hyacinths, Crocuses, Narcissi and Daffodils (above). These looked great of course, but were also a useful source of early spring nectar for bees. It was worth waiting to see what popped up as it would have been easy to dig a lot up by mistake. There's also the risk of 'weeding' getting rid of some useful flowers, so don't be in too much of a hurry to clear things until you know exactly what you have.

HOW TO: GROW THE BEST PLANTS FOR WILDLIFE

Here are ten of my favourite plants for insects that you can grow easily in the garden:

1. Globe thistle
2. *Buddleia davidii*
3. Fennel
4. Lavender
5. Ice Plant
6. Red Valerian
7. Red and White Campions
8. Hawkbits
9. Ox-eye Daisy
10. Foxglove

Globe Thistles are a top plant for your borders – bees love them.

It is important to have good structure to your flowerbeds, and mimicking the vegetation of the surrounding areas makes a natural extension of habitat and will encourage species to extend their ranges into your garden. Have a look and see what flowers work best for insects in the wider area and then grow them yourself.

Go tall at the back of borders and gradually bring down the height as you get to the front. This makes viewing easier for you as well as making maintenance easier. Leave some bare patches of mud and earth for insects such as solitary bees that burrow tunnels. You can even turn parts of your lawn into a meadow if you are feeling bold.

SPRING MIGRATION

Spring migration is perhaps the most anticipated event in nature's calendar and the number of bird species in your area will increase by the day from late March. There are two main sorts of migration in spring: the arrival of birds that have flown from Africa to breed in your area; and birds heading further north in the UK and to northern Europe, and even the Arctic. In the case of the latter they will literally fly straight through and only certain weather conditions will bring them your way, or make them stop.

Diary Notes:

1st May – A deluge of terns

I suddenly found the need to turn on the light while reading this evening. As it was only 7 o'clock, I was a bit puzzled, but all became clear as for the next hour, one of the heaviest deluges I've ever experienced fell from the sky. An hour later, it finally stopped and with half an hour of daylight left, I couldn't resist the urge to see if anything had dropped in at the reservoir.

A quick scan of the eerily dark waters revealed nothing but then, over the far shore, a flock of birds, rising up into the sky before dropping down again. There were a good couple of hundred birds, so I assumed they were Black-headed Gulls arriving to roost. Telescope locked on and I could not believe my eyes. They were terns and Arctic Terns at that. As I watched them skimming the surface, birds started flying the other way through my scope. There were two flocks and I tried to get an accurate count, managing 460 birds. The flocks kept rising up, hoping for clear skies to allow them to continue to migrate, but the gloom set in again and it was time to go. This smashed the country record for this migrant in one of those arrivals that could so easily have happened without anyone knowing anything about it.

Keep your eyes skywards for a Scotland- or Scandinavia-bound Osprey between March and May.

Migrants in spring, compared to autumn, tend not to stay long. They have places to go and breeding to get on with. In some cases, it may just be a day or two's stop to feed up before continuing, but other birds may just be visible for a minute or two as they fly straight through. This is what makes connecting with migrants so exciting. On days when I know migrants are on the move nationally, I visit my local reservoir (a five-minute drive from home) up to three times a day to try and catch all the action. It's very unusual for the same birds to be present on each visit. Early mornings reveal the birds that arrived overnight, but evening visits will bring birds that migrate during the day and have dropped in.

British people like to talk about the weather a lot, but birdwatchers and naturalists like to talk about it even more. Most days in spring, I'll check the weather to see if it will be warm enough to go looking for Adders, sunny enough for butterflies or if the wind direction means it's time to head off looking for migrant birds. Watching the weather is a must for making sure you catch up with as many migrants as you can.

The 'right' weather conditions can

HOW TO: USE THE WEATHER TO FIND SPRING MIGRANTS

If the weather is too calm and sunny, birds migrate straight over and are invisible to the naked eye, so don't wish for too much 'nice' weather in spring!

• Rain forces migrating birds down from on high to large bodies of water. They sit out the storm until the skies clear and they can continue on their journey.

• Flocks of Common Scoters are a frequent sight around the UK's coast, but in spring they make shortcuts across land at night, spending the day sleeping on reservoirs.

• Kittiwakes fly cross-country en route to and from their coastal cliff nesting sites in spring and autumn. In my part of the world, they cut across from the Severn Estuary, heading to the east coast. They can even be seen in flocks between February and April as they head to breeding sites. Strong winds could bring them to your local reservoir.

• The world's smallest gull, the Little Gull, and Black Terns move

Migrant Ring Ouzels love to search for food on areas of short grass.

through in spring. If the wind goes east combined with rain in April and May, head for open water.

• Don't despair if you lack wader habitat in your patch. In spring, they still move through, and lower into a north headwind, so pay close attention to any flocks of birds passing over at height looking like clouds of fast-moving smoke. It's easy to dismiss them as Feral Pigeons, so check every flock.

Look for Black Terns after easterly winds in spring.

literally make the difference between seeing nothing in spring and experiencing flocks of migrant birds pouring through. The weather has a great many local variations, but some general principles will ensure you're primed and ready to go when the weather map has the isobars, wind arrows and clouds in the right position. See the tips on page 91 for the best weather for seeing birds that are heading north either through the UK, or to the east through continental Europe. Generally, even a little wobble in the wind to an easterly direction will bring interesting species to the UK. There are few times of the year when an east wind is not good news for birders.

The weather plays a big part in seeing the many species of wader and tern powering northwards in spring to reach breeding grounds in the high Arctic, often on tundra or far-flung UK locations such as Shetland, Orkney or the Flow Country. One of the most exciting things about seeing birds like these on your patch is speculating where they are going. Is that flock of Whimbrels heading over the garden on its way to Shetland, Orkney or even northern Europe? For just a very short space of time, maybe even minutes, they pass through your part of the world and that is something very, very special.

Diary Notes:

2nd May: To the Arctic and beyond

A bitter north-east wind was in play when I woke up this morning. It's my favourite spring wind as the promise of wader and tern migration up the valley comes with any northerly blow. I wandered outside to scan the valley in case any waders or terns were moving. Seeing a flock of terns beating northwards, rather than the usual leisurely fishing patrols by the local Common Terns, saw me heading in for the scope with high hopes. Three quarters of an hour later, I'd logged half a dozen or more flocks of Arctic Terns – the world's greatest long-distance migrant – making their way through my little part of the world. Realising that I was still in my dressing gown as the curtains were drawn by our neighbours, I made my exit.

As I'd hoped, there were more Arctic Terns at the reservoir after work, but a loud trumpeting call kept drifting across the water. I couldn't work out where it was coming from and then as I checked the 'Mute Swan' on its own in the middle of the reservoir, I realised it was a Whooper Swan. This was pretty late for a Whooper, a rare visitor anyway, and I feared it had lost its mate.

Add spice by keeping a note of your arrival dates for migrants and challenge yourself to beat your earliest each year. Delving deeper, you'll see how weather conditions impact these arrival dates. Cold springs with northerly airflows will see delayed arrivals, but a good southerly airflow originating within Africa will see migrants pouring in. Spring arrivals are stretched out with the first Sand Martin in mid-March, but Swifts and Spotted Flycatchers still arriving in mid- to late May.

Some species are very site-faithful in spring, so remember where you saw them in previous years. Whether it is the same birds returning, the offspring of these birds that have had information on good feeding sites passed on, or just coincidence, it is one of those things that gets you thinking and makes patch watching so interesting. Whatever the reason that species such as Ring Ouzel turn up in the same paddocks, it is well worth remembering those sites, when you saw the birds, and in what conditions you saw them in.

Be alert to the very real possibility of rarities at every turn at this time of year, so think big. Keep picturing finding that Hoopoe on the vicarage lawn or school playing field or that Bee-eater flying over your garden. It happens on someone's

A breeding-plumaged Black-necked Grebe could drop in on spring migration.

patch every single year – and it could happen to you.

The third week of April to the middle of May is peak migration time in my part of the world. Most migrants are well on their way through by the end of May and things start to settle down for the breeding season. Watching your local birds and hot-spots will tell you much about migration in your area and you'll soon know what weather conditions are best and when and where to look for certain species.

SPRINGTIME STARS

Two of the blue-riband migrants for me are Black Tern and Little Gull. The sight of a party of these delicate dip-feeders flying low over the water is the one I hope for most as I scan the waters of the reservoir. Black-hooded adult Little Gulls flash their jet black underwings, while the plain-faced younger birds show black 'W's across their upperwing. Getting the right weather conditions is vital for these birds (see page 91). During optimum conditions, if there is one flock, there are usually more to come later in the day. You can see both species in autumn too, but they are at their best in breeding plumage.

Migrants on land are also looking at their best in spring. Keep an eye on newly-sown fields because Dotterels are

It's easy to work out how many Cuckoos you have on your patch from their song in April and May, but seeing a Cuckoo requires a bit more thought.

• Follow the sound of the calling males – they'll often perch up in a tree – and then wait patiently for them to move.

• Cuckoos look very like Kestrels in flight, so tune in your search engine.

• This nest parasite has regular 'hosts' for its young. Find the hosts and you might be treated to the amazing sight of a Cuckoo being fed by its tiny foster parents. Due to the abundance of wetlands on my patch, Reed Warbler is the main host, but if you are in a downland or upland area, it may be Meadow Pipit.

• The female Cuckoo's bubbling crescendo is less well-known, so tune into that too.

on the move from mid-April to mid-May, stopping off in 'trips' on the way to their mountain breeding grounds. Find and check pea fields, especially, as these have a strange attraction for these land-loving waders. This is a really special bird that could turn up in a field absolutely anywhere. Wheatears like short turf and bare or newly sown fields. Keep an eye on the timings of arrivals in your area. March birds tend to be of the race that breeds in the UK, mainly in upland areas, but April and May bring the larger, more upright and more colourful 'Greenland Wheatears' that migrate as far as Canada. Flocks of Yellow Wagtails are also worth checking for the continental Blue-headed Wagtail in April and early May.

UNDER THE RADAR

The Trans-African migrants take the headlines in spring, but getting to know your local birds well will reveal the movements of a whole set of other species that you might not really associate with being migratory. Common Gulls and Lesser Black-backed Gulls pass through in big numbers in spring, from as early as February. If you watch your local flocks or roost, you'll see a noticeable rise in numbers of both of these species in March and April.

Other migrants can be overlooked as

Normally coastal, Bar-tailed Godwits travel cross-country in late April and early May.

well. I noticed that the number of Ringed Plovers at my local gravel pits leapt up in May. These birds seemed smaller and darker than the one or two pairs that bred there and I realised that in the last three weeks of May and first week of June, flocks of *tundrae* race Ringed Plovers were moving through. It's only though knowing your local birds, their behaviour, their numbers and when they should be there that you can pick out things like this.

Another easy-to-miss spring migrant, best searched for in April, is the White Wagtail. Scrutinise your local Pied Wagtail flocks from mid-March to early May and you should find this smart wagtail standing out from the crowd.

Watch your local Chaffinches for continental birds passing through northwards. These are easy to miss unless you keep an eye on the numbers in your local flocks and your garden. If numbers suddenly shoot up in March and April, it is likely that continental migrants have joined your residents.

INSECTS TAKE TO THE AIR

Birds are joined on the wing in spring by the two most popular families of insects: butterflies and dragonflies. The Brimstone heralds the arrival of spring, but as the days pass, more and more butterflies emerge. One of the most beautiful spring butterflies is the Orange-tip. Lanes and hedgerows are good places to search for it during its peak flight-period in April and May. The females lack the orange tips of the males, but wait for any whites to settle and check the underwing: the female Orange-tip has a beautiful chequered pattern on the underside of its hindwings.

The perfect excuse to take a walk, and do some important conservation work, is to do a butterfly transect (see page 99). I love walking the lanes from home to see what numbers arc like. Doing this each year allows you see how numbers

QUICK TIP: MAKE A CLEAN SWEEP

The first time that I saw a sweep-net in use, I could not believe how much wildlife you can catch by just sweeping it vigorously through long grass and other vegetation. They are easy to obtain from wildlife suppliers but you can make your own from a coathanger and an old pair of tights. It's a great way to study spiders, beetles and all sorts of creatures that spend most of their lives hidden from view.

Diary Notes:

6th May – An early morning stint

There's something about the shuffling, sparrow-sized Temminck's Stint that I've always been a bit obsessed with. I've been checking the records and almost all the Ouse Valley sightings fall between 5th and 25th May. This has been the May I wanted to find my own.

Dawn is coming shockingly early at the moment and it's been a week of half-past-four starts to make sure I'm at the gravel pits before the workers arrive, and before other birders arrive. The main pit looks prime and every day this week the water-level has dropped, each time revealing a few more inches of mud.

I was still bleary-eyed as I set up my scope on the bund this morning and began scrutinising the islands. Two spangled Wood Sandpipers delicately picking flies from the shallow water made it a worthwhile visit, but I was just reaching for the lens cover for my scope when a tinkling call sounded out. I scoured the skyline and saw two tiny waders drop in from the south. Two Temminck's Stints in their breeding best. I couldn't wait to get the news out. A great start to the day and maybe a chance for a lay-in or two for the rest of the week.

are faring. You'll find that there is quite a marked year-to-year variation in your local butterfly populations.

Two of the earliest butterflies on the wing each year are the easily-missed, moth-like Grizzled and Dingy Skippers. They are on the wing from early April and are good targets while you're waiting for the first migrant birds to arrive if passage is slow going. Disused railway line embankments are excellent places to find them and this is a habitat in general that's well worth seeking near you.

In spring woodland, the Speckled Wood is a subtle star. It's unusual in that it is happy to fly in 100 per cent cloud cover, so even on dull days you can at least be assured of some butterfly action.

DRAGONS AND DAMSELS

The dragonfly season starts as early as April with the emergence of the first

The Grizzled Skipper is an early spring species.

Diary Notes:
13th May – Orchids on display
A friend tipped me off about an impressive display of Green-winged Orchids at the racecourse. After work, I headed over and was scratching my head as to how this rare plant could grow on a busy racecourse. Scanning with binoculars, I could see that the entire middle section between the track was awash with pink flowers. I nipped in under the rails and, having failed miserably to estimate the numbers after getting to 2,000 plants, I just sat back and enjoyed the show. This colony is protected, which is fabulous. It is great to see rare things in unusual places.

A Whitethroat was going ballistic with its somersaulting display flight and scratchy song in the Hawthorn right next to where I'd parked. He's either struggling to find a mate or is a late arrival making up for lost time.

tempt them onto your finger at these times. Be gentle and don't force them, but it's a wonderful experience to admire the incredible eyes of a dragonfly, full of thousands of tiny mirrors, at close range.

Another good thing about spring dragonfly watching is that the fresh individuals head to nursery grounds in long vegetation shortly after emerging. Once they are ready, they head to water and set up a territory that they defend furiously. I found a nursery in a sheltered corner of my local water meadow and it's where I see most dragonflies each year and get amazing views. Check sheltered spots close to waterways, especially places which get a lot of sun, and you could have some great dragonfly experiences in spring.

Many insects use the power of the wind to help them disperse. I've noticed that if the wind blows from the east or

damselfly, the Large Red Damselfly, and the first dragonfly, the Hairy Dragonfly. One advantage of dragonfly watching in spring, compared to the hotter conditions that come with summer, is that the lower temperatures make dragonflies less frantic in their movements so it's a good time for getting good views – and for photography. In addition, newly-emerged dragons and damsels have to pump fluid into their saggy transparent wings to make them ready for flight. You can even

Female Scarce Chaser. This species is expanding its range in the southern UK.

south-east in May, arrivals of freshly-emerged, colourless, teneral Common Blue and Azure Damselflies appear in the hedge in our garden. They must use the wind to help them disperse from the direction of the river. Work out where the nearest waterbody is in relation to your garden and keep an eye out for weather that will bring damselflies to you.

It's not long before dragonflies, and a whole lot more wildlife besides, get on with summer's main objective – reproducing. Migration is over, bird song dwindles to nothing and although birds become harder to see for a time, it's the time of year for other groups to come to the fore in a riot of colour.

Diary Notes: 30th May – We have Badgers!

Nine in the evening and Springwatch had finished on TV, filling me with inspiration about the summer wildlife-watching season ahead. With a good hour's daylight left, I went into the garden to gaze over the hedge into the field, hoping to see an owl. In one of the moments when you sense rather than see, I turned my head left to see a low-slung shape, blazing black and white stripes on its head, barrelling up the field edge, before scurrying into the hedge surrounding the garden next to ours. I wanted that garden to be ours more than anything at that moment. We have Badgers.

The next night I hopped over the fence to put out food on the track next to our hedge. I picked up a bag of dog biscuits from the supermarket, a bit worried that my choice of the cheapest range of biscuits might attract some disapproving glances from pet owners.

I eagerly waited, watching those biscuits, but nothing came. The following night, the sound of crunching as soon as I stepped outside saw me approaching the hedge with the lightest footsteps I think I've ever made. I'd done it – a Badger was feeding there. It immediately accepted me watching from eight feet away hidden behind the privet hedge and hardly daring to breathe. It stayed for 10 minutes, enough time to gobble the lot before turning and scurrying off back up the path.

Get to know the habits of your local Kingfishers and you will be enjoying views like this

Colourful insects abound in summer, including the spectacular Small Elephant Hawkmoth.

SUMMER

Whether you are an early bird or an evening stop out, summer extends the day's wildlife watching time considerably at both ends of the day. That alone makes this season a truly wonderful thing. The amount of wildlife on show, wherever you look, is staggering and it's a time of year when I for one resent having to go to work when the sun is shining and the borders are buzzing.

It's the time for frayed nerves as parent birds bring their 'oh-so-cute', and vulnerable, young to the garden, Badger cubs emerge from their setts, and it's the time to get looking at meadows, field margins and roadside verges to put your insect ID skills to the test.

Butterfly numbers are at their peak, spectacular orchids stand proud in the grass and there is no better place to be than messing about by your local river to enjoy wetland wildflowers and dazzling dragonflies and damselflies.

HIDDEN TREASURES

When the rush of spring migration is over, it's the perfect time to take things at a more leisurely pace. Walk more slowly, lay down in a meadow and peer into shallow water to discover some of the thousands of easily-overlooked smaller species. It's a time to finally try and put a name to that beetle you see each summer under the compost bin, the yellow ladybirds in the border and the flies that sit on the lilypads in the pond. Try concentrating on a different family each summer. Anywhere rich in flowers is the perfect spot to start looking.

Beetles are an enormous group in the UK with about 4,000 species in total. They can be very tricky to identify, but the longhorns are big, distinctively marked and colourful – perfect! Look for them in summer, including the Spotted Longhorn along woodland rides, the Wasp Beetle and the superbly named Greater Thorn-tipped Longhorn.

Other beetles to look for in your garden borders include the Swollen-thighed or Thick-legged Flower Beetle. I keep a small patch of lilies in my garden so that the Scarlet Lily Beetle can come and munch on them – it's a sacrifice I am happy to make. Other beetles to watch for are the Devil's Coach Horse, which curls up its rear-end when threatened, the Lesser Stag Beetle and, if you are really lucky, the magnificent dusk-flying Stag Beetle.

Diary Notes:

4th June – A currant affair

The clearwing pheromone lures that I bought to see what species of these lovely little moths are present in the patch have been really successful so far. Red-tipped Clearwings buzzed in to their lure in sallows at the marina, Six-belted Clearwings came to Birdsfoot Trefoil on the short grass at the reservoir's lagoons and Red-belted Clearwings appeared in an old apple tree I found in the churchyard.

Today though, I managed to find a new species without even using them. I was looking out over the valley when I glanced down at the Privet hedge beneath me and was amazed to see a clearwing perched on one of the leaves – not exactly typical clearwing behaviour. It was a black-and-yellow-striped Currant Clearwing. Later in the day I headed off to the garden centre to invest in a blackcurrant and a redcurrant bush in the hope it would lay its eggs there. There are plenty of raspberry canes too, so fingers' crossed for the newly-arrived Raspberry Clearwing.

Remember to look down as you walk along paths in summer because many insects bask, wait for prey or make tunnels on the path. Check dry and heathy spots for the jewel-like Green Tiger Beetle. They flip up at your feet and land several metres away, flashing like little emeralds as they shoot through the air.

There are around 46 species of ladybird in the UK and you could easily find 10 or more species in your garden. The non-native Harlequin Ladybird is causing havoc, but you should still be able to find plenty of natives, from the yellow-and-black 22-spot Ladybird to the familiar and much-loved red-and-black Seven-spot Ladybird. Less familiar species to search for include Eyed, Kidney-spot and, in wetter areas, the Water Ladybird.

Bloody-nosed Beetle – one of 4,000 UK beetle species.

Weevils are another group that is very easy to overlook – they are tiny! Even if you can't identify them with certainty, these whacky looking little insects are well worth studying with a hand lens. The Acorn Weevil has to be seen to be believed with its black eyes and super long proboscis. The Nettle Weevil and Hazel and Birch Leaf-rollers are perfect examples of how knowing the plants insects are on can help you ID them.

There are lots of colourful and easily identified bugs too. Shieldbugs are well named and have the added advantage of being easy to identify from their colours and markings. You should be able to find half a dozen species in your garden border, from the Hairy Shieldbug, which can be very numerous, to Birch, Green, Red-legged, Hawthorn and Parent Shieldbugs.

QUICK TIP: LOOK ON FENCEPOSTS

Never walk past a fencepost in summer without checking what's on top, or on the side. All sorts of 'micro-wildlife' use them, from jumping spiders, robber-flies and weevils to spider-hunting wasps and amazing serpent-necked snake-flies, especially where older wood is used. It can make your walks take a lot longer, but it's well worth the effort.

There are around 270 species of bees in Britain and Ireland and they come in a huge variety of shapes and sizes. If you thought we just had bumblebees and honeybees, you're in for a treat. Your garden is a great place for getting to know many of the families and individual species in summer. One UK garden, looked after by a bee enthusiast, has a bee list of more than 130 species.

Cuckoo bumblebees are parasites in the nests of 'normal' bumblebees. These are tricky to identify but once you get familiar with their less hairy appearance, darker wings, lack of pollen baskets on the hind legs of the females, and their sluggish way of moving when at rest, you could be adding Gypsy, Red-tailed and Vestal Cuckoo Bumblebees to your garden list.

Scorpion Fly – common, but impressive.

QUICK TIP: PEER INTO FLOWERBEDS

Poke your nose into flowerbeds and keep on peering. An astonishing array of bugs, beetles, flies, bees and wasps can be found resting, nectaring and basking on flowers. 'Open topped' species, such as Tansy, Buddleia, Fennel, Globe Thistle and Ox-eye Daisy provide the best viewing, as things perch in full view and are easy to photograph.

Leafcutter bees are responsible for the perfectly formed semicircles of missing leaf and mining bees are a big group of around 70 species that tunnel into lawns, sandy paths and bare soil.

The chirping or reeling of crickets and grasshoppers from long grass is a classic summer sound, but how often do you take the time to track down the source and try and put a name to what's making it? There are around 30 species of crickets, groundhoppers and grasshoppers in the UK and you'll have several living close to you. One species that is rapidly moving north through the UK is the Roesel's Bush Cricket. Have you heard a high-pitched buzzing like an electricity wire in summer? That'll be them.

Long-winged Conehead is another recent arrival. Not only do they have a fantastic name, they are beautiful brown-and-green crickets that are well worth looking for in the long grass.

There are lots of easily identified flies as well. Peering carefully into bushes and hedgerows will bring lots of rewards in summer. A common, but

Hornet Moths emerge from poplars in June and July.

always spectacular, summer insect is the Scorpion Fly. Look for it resting on leaves in hedgerows and woods and in wetland vegetation.

Sawflies are colourful creatures that everyone has seen and possibly commented on what whacky looking insects they are, but might not know their identity. There are some beautiful species, most with just Latin names. The Figwort Sawfly, Birch Sawfly and Large Rose Sawfly are all easy to recognise and ones to watch out for. Umbellifers such as Cow Parsley, Hogweed and Wild Carrot are great spots to search for sawflies, and many other insects, especially when the white flowers are at their peak. Try wetlands as well as woods and roadside verges to increase the number of species you see.

HOW TO: FIND CRICKETS AND GRASSHOPPERS

You'll have a dozen or so species of grasshoppers, crickets and groundhoppers in your local area, but getting a good view of one, especially crickets, is not easy.

• Approach a 'chirping' individual step by step, stopping until it starts again and then slowly move closer.

• Get down low, inch your head forward, stop, go a bit further and gradually you'll get close to the stridulating individual and be eyeball-to-eyeball with one of these fabulous insects.

• Keep an eye on the lowest rungs of fences that run through long grass because these can be favoured perches.

HOVERFLIES

Perhaps the most colourful and attractive of all flies are my favourite group of insects: hoverflies. They perch obligingly for long periods on flowers and other vegetation, and thanks to the arrival of new user-friendly field guides, you will be able to find and identify many of the UK's 280 or more species on your patch. Looking for hoverflies will get you into lots of habitats, including woods (where the abundant *Syrphus* species create a real buzz), hedgerows, meadows and roadside verges, and there are wetland specialists too. Visit them all and your hoverfly list will soon be soaring.

Gardens are among the very best places to get to know hoverflies. I've recorded about 30 species in our young borders and on the lawn where I let hawkbits grow up in late summer. If you have an established garden, you should be able to find even greater variety.

The tiny and very distinctively-marked Marmalade Hoverfly is the most familiar species and will be common in your garden. Many hoverflies are mimics of bumblebees, honeybees, wasps and even hornets, so don't be fooled. *Eristalis tenax* is another species that you will probably see every day of summer, but might not know what it is. It's a honeybee mimic. *Volucella bombylans* is an attractive bumblebee mimic and even comes in red-tailed and white-tailed forms for added disguise.

The yellow-and-black wasp mimics will definitely catch your eye. *Xanthogramma pedissequum* and the *Chrysotoxum* species are superb. Our two largest hoverflies are Hornet mimics and are truly impressive beasts: *Volucella zonaria* and *Volucella inanis*. I have seen more of these on

Buddleia davidii than anywhere else. If you have them in your garden, keep checking the flowers between June and August for these striking hoverflies at rest on the long flowerheads. Not only do they look like Hornets, but they can also mimic their buzzing. They are inactive and a

Meadowsweet is a draw for hoverflies.

Bee-mimicking hoverflies include species in the genus *Eristalis*.

bit sluggish, and therefore provide good photo opportunities.

Look out for another member of this genus, *Volucella pellucens* (the 'Great Pied Hoverfly'), with its black-and-white body. It hovers perfectly still just above head

Volucella inanis is one of two hornet-mimicking hoverflies you could find.

HOW TO: SEE COMMON LIZARDS

These are great little reptiles and can be found in a wide variety of places.

- Look for them basking in the open in summer. Fenceposts and woodpiles are good places to search.

- Approach slowly and tread gently, as they'll soon scurry off back into cover.

- They will return to the same spots, so just wait quietly and you should get a second bite of the cherry.

QUICK TIP: HOLES AND TUNNELS

There are hundreds of species of solitary bee and wasp and many can be found in your garden. Keep an eye on any holes and tunnels that suddenly appear, having been excavated in light and sandy soil and holes in bricks. The ruby-tailed wasps are truly tropical-looking. When the sun lights up their purple and green iridescent colour, they shine like gemstones. Spider-hunting wasps are very entertaining to watch as well.

height along woodland rides in June and July.

The Long Hoverfly (*Sphaerophoria scripta*) is also unmistakable and 'The Footballer' (*Helophilus pendulus*) can be seen on lilies in ponds. It's a regular at my garden pond now in summer. *Xylota segnis* is easy to identify because it scuttles around on leaves a bit like a beetle. The list goes on and on, so get out there and start discovering what hoverflies you have.

DRAGONFLY PEAK SEASON

You'll find the peak number of species, and individuals, of dragonflies and damselflies on your patch between June and August. If you have fresh water – a slow-flowing river, canal, ponds, lakes or gravel pits – you can rack up double figures of species easily at this time.

Our most magnificent dragonfly brings something prehistoric to our summer waterways. The sight of a male Emperor Dragonfly (our biggest species at 8cm long and with an 11cm wingspan) patrolling the water's edge, and stalling right in front with its mirror-laden eyes fixed on you, is something else. They first appear in May and for the next few weeks they rule the waterways with their regal presence. Find a waterside spot, sit down and let them do their stuff before you.

Diary Notes:

12th June – Marvellous mammals

The wheat is still only a few inches high, so every evening I've been scanning the field from the garden. A Muntjac emerges on the bank every night to stand and stare before walking out into the field to graze.

Brown Hares are a regular sight as well with up to three careering across the field. The stars of the show have been the Badgers though. Tonight, I watched the adults gathering bedding and comically scuttling backwards to drag it into the sett that I've now managed to pinpoint. They are so easy to see right now, so needless to say Laura is seeing a bit less of me at the moment.

With the Barn and Little Owls also appearing regularly at dusk, plus Noctule Bats feeding high over the trees growing behind the bank and Pipistrelles darting around my head, it's pretty spectacular viewing in the evenings at the moment.

Watch for Emerald Damselflies in rushes and reeds by still water.

One of summer's most striking hoverflies: *Volucella pellucens*, or the 'Great Pied Hoverfly'.

HOW TO: WATCH FISH

As a child, I used to love staring at shoals of Rudd and Roach beneath bridges. There is something very relaxing about standing on a bridge and watching fish swimming in the clear waters beneath you.

• A pair of close-focusing binoculars will enable you to identify several species and provide great viewing.

• The peak viewing period is when the water is clear in summer.

• You could easily find more than 10 species in your local river, so find a spot on a quieter riverbank and track down any bridges.

• Scan under overhanging vegetation for species such as Tench.

This female Southern Hawker laid eggs in my garden pond.

Some bird species are poorly named – take the Garden Warbler (have you ever seen one in a garden?) – but it's a different story with our dragonflies. They are conveniently named after the way that they fly. Darters dart out from a fixed perch, returning to the same one after a sortie. Hawkers do just that, flying up and down with precision along the same woodland ride or hedgerow. Chasers are the never-stop species that whizz around in a flurry of colour low over the water.

Take a good look along your local waterways in summer for the Banded Demoiselle damselfly – a stunning creature that would not look out of place in the Amazon. As they fly up from the waterside vegetation, usually in good numbers, they flash their deep blue wing-patches. Especially if you live in the west of the UK, you may have its close relative the Beautiful Demoiselle, which prefers moving water.

Any new ponds, either created by you, or as an accidental by-product of industry or development, are top places to search

---

Diary Notes: 18th June – Hornet Moths emerge

A winter of reconnaissance finally paid off early this morning. At the marina I'd found at least 40 old poplar trees – the habitat of the spectacular Hornet Moth.

I took an early morning visit to see if I could catch any newly-emerged adults today. It didn't take long to find tell-tale emergence holes at the base of several of the trees and some fresh larval cases. Thirty-eight trees later, and more than one strange look from the local dog walkers, and my trunk-inspecting gaze was finally stopped by a bright yellow chunky body clamped to the trunk. What a moth this is! Incredibly, by the time I had walked back its wings had barely hardened, but it had attracted the attentions of a male and the pair were copulating. Unbelievable to think that the pheromones are so strong that it could attract a male just minutes after emerging.

The meadow is still flooded and I got a real treat there, finding a family of six Foxes sleeping and playing on a muddy reed-fringed spit. There had also been an emergence of Scarce Chasers, still wonderfully golden-orange before the males turn powder blue. There were dozens resting in the sedges and rushes around the meadow. I was able to pick one up that had got into a spot of bother. Four Hairy Dragonflies were also busy patrolling along the Hawthorn hedges.

Watch out for our largest snake, the Grass Snake, in and by ponds.

in late May and June for the chunky Broad-bodied Chaser. It is usually the first dragonfly to colonise new ponds.

Southern Hawkers have an endearing habit that you can take advantage of to secure great views. They are inquisitive and will fly up to you to check you out. Find one patrolling a territory, position yourself on the flight path, stand still and you'll have one of these big beauties staring at you from a few inches away. This is one for the July and August 'to do' list as this is when they are at their commonest.

While you are dragonfly watching, also keep an eye out for our largest snake. The Grass Snake preys on amphibians and can swim very well. If it's a still day and you suddenly see a strange ripple moving across the surface in jerky fashion, you might well have found one.

BUILD IT AND THEY'LL COME

Summer is the time for enjoying the fruits of your labours and the sweet scent of flowers, but make sure you have time to create, and then keep an eye on, the smelliest part of your garden. Composting is a great way to recycle your food and garden waste and heaps provide a fantastic warm micro-climate for all sorts of creatures. Our only legless lizard, the Slow-worm, loves compost heaps, as do White-legged Millipedes, Brown-lipped Snails and a host of other minibeasts.

All sorts of surprising items can attract more wildlife. One of my favourite groups of insects is the solitary wasps. These are much less feared than the familiar yellow-and-black social wasps and come in a huge variety and number of species. They are not easy to identify but that doesn't make them any less interesting. Garden borders can be home to an impressive array. One bonus that came from me 'edging' our flowerbeds with the old bricks was that I used the types with holes in them. These are now being rapidly colonised by several species of solitary wasps taking advantage of the ready-made tunnels.

Get your hands on any squares of carpet that might be going spare and any old tins or sheets of corrugated iron. Place them in your garden and you might be rewarded with a Common Lizard, Slow-worm, or maybe even an Adder, sheltering under them.

One of the most rewarding things about making a pond (see the 'Spring' chapter) is attracting dragonflies to your garden, but the next step is to have them breeding there in summer, so provide

Let your lawn go wild for a few weeks in summer and make a mini-meadow.

Broad-leaved Everlasting Pea is good for bees.

plenty of tall vegetation around the edges. The larvae of some species spend several years underwater as voracious predators.

I always think that the best time for butterfly viewing in the garden is July and August because the smell of the buddleia combined with the sight of these butter-flies is one of summer's highlights. A good sized 'butterfly bush' will attract dozens of butterflies crowding onto the flowers, with Small Tortoiseshells and Peacocks providing stunning colours. Don't miss it.

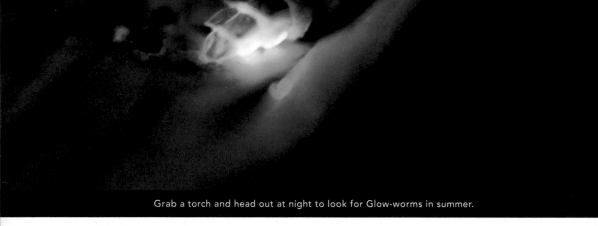

Grab a torch and head out at night to look for Glow-worms in summer.

GAME, SETT AND MATCH

I've always enjoyed wildlife-watching in the twilight hours, so it's lucky that one of my favourite animals comes out at this time of day. This one species has been instrumental in my obsession with my local wildlife. The iconic Badger is not rare in the UK, but before we moved to our new home I had only seen them a handful of times, mainly at a specially constructed Badger-watching hide in an old orchard where they are fed under floodlights. Little did I know that I'd be creating my own viewing area in my own garden. In a classic example of getting to know your local area and what lives there really well, I now see Badgers every summer evening and enjoy hours of intimate viewing each year – and so can you.

Once I'd glimpsed my first Badger, it was all systems go to make them a regular sight in the garden. I was amazed how easy it was to tempt them in – and to keep them coming back. Once they felt safe enough to come to the food I put out, they would come every night and they are remarkably consistent in their arrival time. On one memorable June evening, eight Badgers were present at once. I even managed to get them to feed on the back doorstep for a few weeks one year. Being nose-to-nose with a Badger with only a pane of glass separating you is something very special.

Badger enjoying dog biscuits on my doorstep.

HOW TO: WATCH BADGERS

• Start your search in winter and early spring by tracking down setts when there is little obscuring cover. Also look out for fur stuck on barbed wire fences.

• Find suitable, and comfortable, spots for viewing. You need to be far enough back but have a clear view, even when the vegetation grows up in late summer.

• Check the wind direction and choose a spot where it isn't blowing your scent into the sett.

• Be in position a good hour before dusk and use binoculars, watching from a sensible distance of at least 10 metres. Badgers emerge in surprisingly good light in summer so don't be caught out.

• Use peanut butter and dog or cat biscuits to try and tempt Badgers into your garden, or to a quiet, secluded spot where you can watch them.

The chances are that there are several Badger setts in your area, so you certainly don't need to attract them to your garden and can enjoy them out in a local wood. Spend some time looking for setts and then you can weigh up the opportunities for viewing and put your plan into place (see left).

I can watch my Badgers feeding on the other side of the hedge, comfortably from a distance of a few feet, but if I am carrying an unusual scent (say if I have been cooking, or the time I tried a new aftershave) they soon bolt. Their sense of smell is epic and they are creatures of habit. If anything seems different from the previous night, they will change their behaviour, or choose not to linger. It is possible for wild creatures to become accustomed to you, particularly if you stick to a routine. They'll simply get on with theirs to fit in with yours.

When you watch the same Badgers

Put in the time and you could get up close and personal with your local Badgers.

on a day-to-day basis, you can get to know individual behaviour, markings and even their character. It becomes easy to distinguish the muscly, chunky-headed male Badgers from the more delicate females, and of course know the super-cute youngsters when they make their first appearances above ground at around three months old.

During one year of particularly prolonged Badger viewing, I named every Badger from our sett based on features such as the colour and shape of their tails. Our original female 'Ginger' was easy to recognise. She's a small badger, but her wispy tail is ginger in colour meaning that even a back-end view is enough to know she's around.

Badgers don't actually hibernate. They will spend much longer in their under-ground setts in the short days of winter and rarely come above ground, but there is a long period when it is possible to see them. You'll find that there is a peak time for sightings of your local Badgers based on several factors, including the speed of growth and height of the vegetation. May to July is the best time on my patch.

The quiet and patient watching and waiting you'll need to see Badgers will also provide you with a wealth of bonus sightings, from chance encounters with owls, to perhaps a Stoat or Weasel popping up on the path in front of you, and of course crepuscular wildlife from Woodcocks and bats to chafers and Stag Beetles. Brown Hawker dragonflies stay out until dusk in late summer and you might be lucky to catch some of the bigger moths coming to nectar, such as Elephant Hawkmoths at honeysuckle or Privet Hawkmoths at privet. Seeing these big moths hovering at delicate flower-heads is a rare treat, so always have a torch handy.

Badger cubs are a summer treat, but the young of lots of other creatures provide great viewing too. One of my most memorable summer days was when a mass movement of 'toadlets' occurred on a warm summer's day after a wet night. They were even getting into local offices, causing some shrieking from surprised employees who perhaps suspected a plague.

BABY BIRD SEASON

Summer is the time for watching and seeing how your local breeding birds have done. So often, success is linked to the weather. If it's a cold, wet spring and summer, the breeding season is usually poor, but a hot one with plenty of insect food that allows the adults to get into peak breeding condition can see a baby boom.

Keep a record of all the successful pairs you find and how many broods the same pair of birds has had. Blackbirds, for

HOW TO: CREATE A WILDLIFE-FRIENDLY LAWN

Mowing the lawn is a job that few people enjoy, but the good news is that you can make a great haven for insects and small mammals by leaving a patch of your lawn to grow long.

• Hawkbit grows in profusion in late summer on my slightly scruffy lawns, but it can teem with hoverflies and a host of bees, so let it be for a week or two.

• In late winter and early spring, Sweet Violets are a feature of the front lawn and they provide early nectar – as well as looking fabulous.

• You don't have to let your entire lawn grow long. Just cordon off an area and let it grow. You can also plant areas with wildflower seeds in spring or autumn.

example, can easily raise three broods in a season. Mute Swan cygnets and impossibly cute Moorhen and Coot chicks are well worth looking for around your waterways and Black-headed Gull colonies are a cacophony of noise and activity once the young have hatched. Common Terns probably take the prize for the cutest baby though.

There are even unexpected experiences to be had in summer as there are always some birds that surprise you. Checking the bigger local gardens with tall trees has revealed a small summer population of three or four pairs of Spotted Flycatcher in the neighbourhood. July and August are good months for locating family parties so tune in to their thin clicking calls coming from the treetops and watch for darting fly-catching sorties into the air.

It's worth putting up an open-fronted nestbox in your garden because Spotted Flycatchers take readily to them. Put one partially concealed in Ivy on a wall, fence or tree, and you could get lucky. The day when Blue and Great Tits leave your nestbox is always exciting and seeing them make their first flight is a summer challenge. Mine are usually gone before I've had a chance to see how many young were born.

In the heat of the day Swifts range widely and rise high in the sky, but on summer evenings they gather prior to

Diary Notes:

20th July – Eggs-citing news

A splendid female Southern Hawker was patrolling the pond this afternoon. It was great to see the work I've been doing on what was once a vegetation-less fish pond paying off. Going local to source reeds, Water Plantain and Water Violet has really paid off.

Things got even better, though, when the dragonfly visitor started to skim lower and lower over the water and perched. Would it? It certainly did – it spent a good hour laying eggs in the pond. Amazing to think the larvae once hatched will spend several years underwater before emerging, breathing air and becoming the next generation of adult hawkers.

Summer brings a baby bird boom –
Woodpigeons are prolific breeders in my garden!

Your local Blackbirds could raise three broods each year.

Big brood – Great Tits find a well positioned, well-made nestbox hard to resist.

roosting. Sit back on a lounger or deck-chair and enjoy the sight and sound of the 'devil bird' drifting around high in the sky or racing down your local streets.

Allow Swifts to nest in the space below your roof and you'll be treated to the sight of the world's most aerial bird making its few annual visits to solid ground and know you're helping a species that's in decline. You can also spend some time, perhaps even combining it with a drink in the local pub garden, tracking down nesting sites in your town, village or city. Suddenly a Swift will break off and zip up into eaves of old buildings – a great sight. Swifts are one of those species that are gone before you know it, so mark up some quality swift watching time on your calendar in June and July. Keep a count and you'll see them start to decline in number and drift off Africa-bound as August goes on.

It pays to keep a close eye on man-made structures, with the possibility

HOW TO:
FIND ALL THE UK'S OWLS

You can find all five species of regular UK owls on your patch. Each requires a different bit of fieldcraft.

• Tawny Owl. Use your ears. The young's high-pitched calls are a giveaway from May. The sight of a fluffy chick perched in a tree is hard to beat for 'ahh' factor.

• Barn Owl. Track down where they are nesting (an old hollow tree, nestbox, or old barn) and you can be sure of hours of action as they come and go delivering voles and other small mammals to the young.

• Unless you live in the uplands, Short-eared Owl will be a winter visitor. The secret to seeing this golden-eyed beauty is to search on still days (they don't like wind and remain hunkered down until hunting conditions improve) and in the last two hours of daylight.

• Little Owl. This blackbird-sized owl is very obvious when perched on top of a telegraph pole by the roadside or on a fence post at dusk. Taking an evening drive around suitable farmland is one of the best ways to find this fierce-looking owl.

• Long-eared Owls are the most

A Long-eared Owl is always a good find.

difficult. Their mating calls are much easier to overlook than those of the Tawny so tune in to the mournful 'hooo' but also the very distinctive 'squeaky gate' calls of the young birds. May is a good month to listen. The adults' wing-clapping displays can also be heard in early spring.

Keep an eye on Swift numbers in July. Numbers will swell as the young are born,

Collared Doves are a confiding garden resident, even when they have young.

of breeding Peregrines on tall buildings, Black Redstarts on old industrial sites, and tiny Sand Martins buzzing in and out of their colonies in old piles of sand at gravel extraction sites.

Another summer star that comes with a good population of Swifts, Swallows, martins and dragonflies is the dashing Hobby. This falcon was, until fairly recently, restricted to the heaths of southern England but its range now extends right through England and Wales and into Scotland. Once the young hatch in late summer and early autumn, the adults will appear more frequently, perhaps hunting over your garden (see 'Autumn' chapter).

Bird migration isn't over in summer. It never really stops, just slows down. Swifts move in reaction to the weather and can travel hundreds of miles in search of better feeding opportunities. Look out for groups passing over or for large numbers gathered over lakes and

Adult Green Woodpeckers are accompanied by their young (left) in summer – families love the lawns in my road.

reservoirs. They may be birds from far afield that have moved ahead of a storm or in search of good feeding.

There can be summer irruptions of birds and one of our most primitive but exciting species falls into this category. If you have pines near you keep an eye, and especially an ear, out for flocks of Crossbills. They are nomadic birds so even if they do not breed near you, summer is a good time to tune in to their clipped, metallic 'chipping' calls. They are quite unlike any other bird call in the UK.

If you find a Crossbill flock, watch for them stripping pine or larch cones down as they hold them in their feet, like parrots, or hanging upside down in the treetops. If they have arrived in your area, a tip for good viewing is to seek out any standing water. They are thirsty birds from all that seed-eating (chew on a pine cone, or a needle, and you'll get a feel for

Grow Teasels to attract bees.

Marbled Whites are on the wing
in June and early July.

Check grassland and verges for
Bee Orchids in June and July.

what it's like) and will come down to drink regularly. Try putting down a dustbin lid filled with water on a quiet ride and see if they come down.

FLOWER FESTIVAL

Wildflowers are not only one of the largest and most diverse families, they are one of the most useful to get to know when it comes to finding and understanding the wildlife on your doorstep. Plants can tell you a lot about conditions such as soil type (sandy or clay, for example) and many are very localised in where they grow. Even simple things like knowing which ones produce good seed crops will help you to track down autumn and winter finch and bunting flocks.

A good starting point is learning the major flower families – forget-me-nots, orchids, St John's worts, clovers, campions and willowherbs to name just a few – so you can at least pin down the right family. Start with learning the commonest members of each family too, so that you have a benchmark by which to compare unfamiliar new species and will know if something looks a bit different. Timing is also key. Cow Parsley is the commonest of the many and confusing umbellifers, but it flowers before many of the other species and dominates verges in May.

A good way to build up your knowledge

Summer skippers include the Essex Skipper.

Look for the eyecatching Elephant Hawkmoth caterpillar on willowherb, fuschia and bedstraw from July to September.

of flowers is by getting to know a new species every time you go out on a walk. By the end of the year, you'll be surprised how it adds up.

Many insects are highly specialised both in terms of the flowers they feed on, but also where they lay their eggs, or what their caterpillars will feed on, so find their host and you are half way there to finding them. Find the plant, find the insect. Easy!

Some species have very precise habitat requirements, for example Hornet Moths need poplars and Purple Emperors need mature woodland with oaks and sallow. Some species complete their entire life-cycle on just one species of plant.

Check Ragwort for the black-and-yellow striped caterpillars of the Cinnabar Moth and look out for the day-flying red-and-grey adults. Some moth cater-pillars are even more impressive to look at than the adults, so search for them in summer. The Puss Moth and Elephant Hawkmoth are great as adults, but also as caterpillars. It was my Dad finding one of the latter in our garden that spurred my whole interest in nature. Spend some time checking Rosebay Willowherb in summer for the amazing 'eyed' caterpillars.

For me the Bee Orchid reigns supreme among wildflowers, and finding a new colony on my patch in summer is a real

Check around lights for moths. Here, an Elephant Hawkmoth is flanked by Poplar Hawkmoths on my house.

thrill. The flowers mimic a female bee so will attract males. It can be found in many habitats, from sensitively managed roadside verges to grassland and the edges of gravel pits. It is such a perfect form that you must make a note in your wildlife calendar to spend time searching for it in June and July.

My local population of Greater Butterfly Orchid is restricted to a single, caged individual in a woodland in my patch, but hopefully you'll find more. Pyramidal Orchids are one for areas of short turf, but perhaps the most likely species to find on your patch is the Common Spotted Orchid. It can be very common in a surprising variety of grassy habitats. The black-spotted leaves are a giveaway, even when it isn't in flower.

Another special group to look out for, even in town, is the broomrapes. These are parasitic plants that appear to be dead! Each species has its own host and some are very rare, but that shouldn't stop you looking. I had never seen Ivy Broomrape, but on a visit to some friends in Bristol, I was amazed to see it growing in profusion underneath the road sign in their urban street, and in every other street in the neighbourhood. This is proof, if it was needed, that every patch has its own specialities.

Don't forget to keep an eye on the flora of your local aquatic habitats, which can be great at this time of year.

It also gets you out encountering various insects that are associated with wetland plants, including a surprising variety of leafhoppers and snails. Ragged Robin, Soapwort, Yellow Flag, Marsh Marigold and Purple Loosestrife will all bring colour to your local wetlands in summer.

Encourage Foxgloves to grow in the garden. They are easy to transplant, so you can soon have a vibrant bed full of these elegant flowers growing where you want them. Long-tongued bumblebees love them. Borage is another must-have: Honeybees cover its abundant droopy bright blue flowers in summer.

NIGHT FLIGHTS

I really recommend that you invest in a special moth trap for your garden. This is essentially a box, that you fill with eggboxes for the moths to cling to, with a mercury vapour lightbulb that you leave on overnight and the necessary electrics. You can buy these ready-made and good to go. Moths are one of the most unappreciated groups of insects in the UK and the variety and number of species is staggering. The 'macro moths' total around 900 species and the much smaller and harder to identify 'micro moths' a further 1,600.

A friend of a friend makes moth traps, so I purchased one from him and couldn't wait to get it out in the garden on the first sultry night. The first time I peered in, I would have been grateful to see just a few small brown jobs to be honest, knowing that the trap had worked, but as I gently took off the lid and lifted the first eggbox, something amazing came into view: our biggest resident hawk moth, the Privet Hawkmoth. There were three in total.

Opening a moth trap on a summer's morning (get there before hungry birds do) is like opening a treasure chest. You just don't know what will be in there as you lift off the lid and turn over the eggboxes. Moth trapping will provide you with a wealth of species from chunky, colourful hawkmoths that will grip incredibly well onto your fingers with barbed feet to those challenging 'micro moths', some of which are also rather neat. You don't have to identify them all, but it's fun trying and as with anything it is best to try and learn the common species so that the rarer visitors will stand out.

Gardens are great places for moths and if you trap regularly, you'll soon have hundreds of species on your garden list. If you get a portable trap, you can take it into other habitats such as woods where you'll increase your opportunities further. That said, you'd be surprised how far moths stray with wetland and

The UK's biggest resident moth, the Privet Hawkmoth, occurs in gardens.

Day-flying moths include the pretty Common Crimson-and-gold Moth.

Perfect camouflage – the Buff-tip moth resembles a birch twig.

HOW TO: FIND MOTHS

Keen moth-ers who put out a trap have garden lists of more than 500 species, so it is well worth investing in a moth trap to see what's visiting your garden after dark.

• The best nights for moths are overcast ones, where the temperature stays up.

• Following sultry summer nights, check out lights around your house and other buildings. And put out your moth trap if you have one.

• Even if there is a promise of rain, which is likely on such nights, you can put a plastic 'guard' on a tripod over the bulb to keep it safe.

• Get up early the next morning, before the birds, so you can extract the catch safely.

Gatekeepers abound in hedgerows in July and August.

woodland species arriving in your garden even if you are a mile or two from the nearest habitat.

Moths aren't just a nocturnal attraction. As you work in your garden, look out for Large Yellow Underwings disturbed from cover, flashing their yellow-orange underwings as they spiral up and away. The Red Underwing is bigger and even more impressive in flight. Then there are the day-flyers such as Six-spot Burnet, Silver-Y and the pièce de résistance,

the Hummingbird Hawkmoth, which I have almost exclusively only ever seen in gardens.

Summer is the peak season for the moths' more familiar day-flying relatives, the butterflies. There are only around 60 species of butterfly in the UK, compared with the hundreds of species of moths. Once summer arrives start looking for the host of new species that take to the wing: Gatekeeper in the hedgerows, skippers on the verges and grasslands, Marbled

Green Hairstreak can be seen in scrubby areas, with peak numbers in May and early June.

HOW TO: SEE HAIRSTREAK BUTTERFLIES

When searching for the 60 or so varieties of UK butterflies, the five tiny hairstreak species pose one of the greatest challenges. The effort and patience required really pays off as one settles in view on a leaf to show the delicate underwing detail before spreading to show the upperwing colour.

• Get yourself in a position where you are not directly below the crown of a tree, but can see a fair bit of it and watch with your naked eye. If the hairstreaks are there, you'll get a glimpse, with patience, of a small, rapidly flying butterfly.

• Once you know where they are settling, scan the uppersides of leaves with binoculars.

• Other flying insects, such as bees and wasps, can do you a favour because they can disturb the basking butterflies as they fly past, or land near them.

• Search out elms for the White-letter Hairstreak. This species can occur in surprisingly small hedgerows containing elm suckers (now that mature elm trees are much rarer).

• The Purple Hairstreak is the most common member of the family. It relies on mature oaks, so scan those in your local wood.

I 'phone-photographed' this Purple Emperor as it took up minerals from excrement.

27th July – Butterfly bonanza

I'd heard rumours that the northward spread of the Silver-washed Fritillary had brought the species to Cambridgeshire, so with temperatures hitting 20°C, I went out mid-morning to the wood, heading straight for the widest, most flower-rich ride. It didn't take long for a brilliant orange butterfly to sweep down the main ride and soon I'd seen 13 with bramble flowers proving a draw and allowing a few prolonged views of this cracking butterfly at rest. A White Admiral took advantage too, and Purple Hairstreaks seemed unusually easy to see in the canopies of the Pedunculate Oaks, with seven flights seen above the canopy. Fifteen species of butterfly were on show today, with some good hoverflies too.

Back home, the unkempt lawn did the business again with a superb orange-studded Brown Argus nectaring on the hawkbits outside the kitchen window accompanied by a couple of Common Blues. Scarcities aside, the highlight of the day was the cloud of Green-veined Whites over the set-aside. Looking from the garden, I estimated four figures easily.

White on downland, and Meadow Brown in the meadows.

Summer evenings are the time to look for another nocturnal flyer – bats. There are 18 species in the UK and there will probably be more species in your area than you think. A bat detector is necessary for firm IDs, but for some species, such as Brown Long-eared Bat, identification is possible in the field. The slow, 'rowing' flight of the Noctule Bat is a treat from my garden each summer and early autumn. Investing in a bat detector is something I'm planning to do because I strongly suspect there are several other species present.

Pipistrelle bats come to hunt for moths and other night-flying insects around houses, so you can even watch them from the comfort of your home. Head down to your local waterbodies at dusk and you can find Daubenton's Bats flying over the water, spiralling around catching flying aquatic insects. They are not rare, but easily missed unless you look in the right place at the right time. I was surprised at how common they were once I started visiting my local gravel pits at dusk.

Put up bat boxes – they have a slit in the bottom rather than a hole in the front, like bird boxes – and allow them to roost in your roof space. You'll be doing your bit for these fascinating and threatened

Migrating Greenshanks could stop off at your local wetland from late summer.

mammals. Watching bats flitting in the twilight, and pirouetting after moths, is a wonderful way to spend the last glimmers of light on a summer's evening.

As the nights eventually start to draw in, a chill in the air one evening tells you that autumn is on its way. Millions of birds, having raised their young, are poised to start heading south and who knows what will be coming your way...

Diary Notes:

1st August – Feeding time

During the weekly chat with Mum on the phone this evening, I was idly watching the Summer Chafers buzzing around the tops of the trees growing on the green outside the study window. Suddenly, a sharp-winged 'missile' homed in and swept through the treetops. Then there were two – the pair of Hobbies from the nest by the meadow had been attracted in by the mass emergence. It was spectacular to watch the pair snatching up the chunky chafers in their talons at will. Presumably they were gathering for their young, as well as themselves. They were back the next evening, but not again, perhaps having found food elsewhere.

Diary Notes:

8th August – A hoverfly invasion

I'm trying not to get too annoyed at the encroaching trees and scrub in next door's garden, because from a wildlife point of view, it's a bonus. A good crop of Ivy has consumed the flimsy, fast-rotting back fence, but arriving back from a few days away revealed a tremendous arrival of hoverflies on the south-facing clump. There were three migratory species – Eupeodes corollae, Eupeodes luniger and Scaeva pyrastri – all presumably carried here by the south-east wind that's bathing us in warm air at present. Dozens of honeybees were getting their fill from the ivy flowers as well and there is a real buzz in the air.

Hedgehogs are very active in autumn, feeding up for hibernation.

AUTUMN

As flowers and insects fade away and autumn's slow decay sets in, it's easy to think that nature slips into a state of inactivity. Far from it.

Bird migration peaks in September and October and autumn brings vagrants from all corners of the globe, plus big arrivals of birds on your patch. Rare moths can also occur, with vagrants arriving from as far afield as North Africa. What better way to start the day than finding a Death's Head Hawkmoth in your moth trap?

A burst of colour comes from a spectacular variety of fungi and the autumn fruits and leaves. Mammals are busy fattening up ready for hibernation and a new cast of autumn insects take the next shift. A quiet time of year? Definitely not!

FUNGI FORAYS

Topping the bill among the next shift of stars in the wildlife calendar are a fascinating, beautiful and challenging group of some 16,000 species with phenomenal lifecycles. Autumn is the season for a treasure trove of treats literally popping up all around you: fungi.

Think of toadstools and the white-spotted scarlet Fly Agaric is one of the first species that probably comes to mind. It's a birch specialist and a truly beautiful find, but the variety of fungi that you can find on your patch is enormous. Hundreds of species produce their fruiting bodies from September into December (in fact you can find fungi year round) – they are exciting to hunt for and pleasing on the eye. There is something thrillingly primeval about the hunter-gatherer instinct, eyes scanning the woodland floor and suddenly discovering the rich purple of a patch of Amethyst Deceivers among the leaf litter, or a chunky Penny Bun (Cep) protruding from the side of the path.

Diary Notes:

13th August – A Scandinavian visitor

I heard it first – a gentle, slurred 'hoo-eet' coming from the apple tree next door. "Doesn't sound right for a Chiffchaff..." was my first instinct. I crept up to the fence and as I did it flew onto it, just above my head, red tail blazing and big black beady eye staring back at me. A Redstart, and another new species for the garden bird list. It spent the day commuting between next door's garden and our fence, snapping up the abundance of flies around at the moment. At one point I could even see it from the lounge. The next day it had moved onto the bramble and elder scrub on the edge of the field, 100 metres away, but still viewable with binoculars from the garden. A Reed Warbler also decided to show up in the hedge, which is a far cry from its typical breeding habitat. By day three the Redstart had gone – next stop Africa.

Dead Man's Fingers – one of many wonderfully named fungi.

Look for Fly Agarics in mixed woodland in autumn.

With regular forays and careful searching, your local park or wood will provide you with the richest variety of fungi, but the average medium-sized garden will also host a few dozen species. Search lawns for waxcaps – brightly coloured toadstools with slimy caps – and check your log pile for species such as Jelly Ear and Common Eyelash. Keep an eye on woodchips and mulch on flower beds for cup fungi. Shaggy Ink Caps (which were once used for ink) and Pavement Mushrooms can even push up paving stones – an amazing feat of endurance that you can see in your local streets.

Fruiting bodies of fungi can literally pop up overnight so even if you think you've already checked an area, every day could bring a new species. Don't be frustrated if you can't put a name to everything you see. Many fungi need microscopic examination for a firm ID, but you should still be able to name a great many species or at least get them to family name. Every fungi season you get under your belt will keep your list growing.

I've been a fungi fanatic since my eyes were opened by the enormous number and variety of species that I found on my patch. Some of my most exciting finds have been fungi, from tiny, perfectly-formed Field Bird's Nests growing in the veg patch to beefy boletes in the woods and alien-looking Palamino Cups that popped up in profusion on an area of woodchippings at my gym. Once you start looking for fungi, you won't be able to stop and you'll be eagerly scanning

Diary Notes:

31st August – Clouded Yellows arrive

Reports of large numbers of Clouded Yellows in southern England have been filtering through. Trying to add one to the garden list by peering out over the field failed, so I jumped the fence and headed out to the old reservoir beyond the Badger sett. I'd only walked 100 yards when a brilliant butter-yellow sprite flew along the track at characteristic rapid speed. I looked back to see if I'd be able to get it on the garden list but it powered northwards with no chance of it lingering. Once I got to the old reservoir, though, I found four different Clouded Yellows dashing over the sandy ground, stopping to nectar at the plentiful dandelions.

The brood of two young Buzzards were still 'mewing' away in the trees on the bank. They look really smart compared to the worn and tatty parents, easily identified by the gaps in their wings and tails.

Earthstars are an exciting fungi find in autumn – this is the widespread Collared Earthstar.

any dead wood you find for the golden prongs of Yellow Stagshorn and Green Elf Cups, marvelling at the shocking blue discolouration of the Inkstain Bolete and searching meadows for the football-sized Giant Puffball.

You could be forgiven for thinking the aliens have landed when a patch of earthstars suddenly crops up. If you find one, give it a gentle squeeze and you can see the spore dispersal as clouds of millions of them puff into the air. The Collared Earthstar is the most likely to be found, probably at the side of a path through your local wood, but there are several species.

Use your nose as well as your eyes when fungi foraying. A sudden sickly-sweet smell will reveal the presence of stinkhorns: Common and Dog Stinkhorns. They are amazing-looking species that never fail to raise a smile when I see them!

Woods will provide you with the greatest variety of fungi. Getting to know

Dog Stinkhorns emit a sickly-sweet scent.

Diary Notes:

19th September: One good tern

I was off to Warwickshire to trade in my old telescope for a new model and decided to give my old one a last run out, stopping at the reservoir en route. The overnight mist was just clearing, as a faint south-easterly breeze blew at my back. I set up at my usual watchpoint in the south-west corner and was soon struck by the presence of a group of jet black ducks on the water – Common Scoters, three drakes and a female. The classic south-east winds had done the trick again, grounding these night-migrants as they cut across land heading from coast to coast.

A few scans later I was stopped by the sight of a tern sitting on one of the buoys in the mouth of the creek, a good half mile away. All the Common Terns had left over a week ago, so I had a hunch this could be something different. Eventually it took flight and as suspected it was a marsh tern that dipped for food, delicately picking insects off the surface. The all-white rump' revealed it was a White-winged Black Tern – only my third ever on the patch and easily the best yet.

which fungi are associated with each tree will open the door to many more finds, and successful identifications, so be sure that you know your local Beech from a Birch and Alder from an Ash. A couple of easy ones to start with are the Birch Bracket that is not only obvious and easy to recognise, but it only grows on birch, and Beefsteak Fungus which is found on oaks.

Beech woodland is a riot of colour in autumn and you should check mature Beech trees for the delightful Porcelain Fungus growing from dead limbs and the Magpie Inkcap in the litter beneath.

Many fungi are poisonous and some deadly, so only pick them for eating if you have them confirmed by an expert you trust. I have a colony of Death Caps growing in the same spot in a local wood. It is a sobering thought to realise that just a capful could kill a human.

Parasol mushrooms are a striking sight.

Autumn is a time for taking advantage of the rich crop of fruits and berries yourself. Blackberries, elderberries, apples, sweet chestnuts and hazelnuts are all well worth tracking down and taking your share of in autumn. Find the trees earlier in the year to make it an easy harvest.

RETURN PASSAGE

At the same time that fungi are cropping up everywhere, the return of wintering birds, and the widespread migration of up to 100 different species, is also happening all around you. Some of it will go on out of sight in the skies well above your patch, but it is the thrills and spills of the migrants and the migration you can see that make autumn so exciting.

It is worth a few early starts in late September and October so that you can see new arrivals passing over in the

Check every bird you see in autumn as you could find a scarce migrant such as a Hawfinch.

mornings: a phenomenon known as visible migration. Some movements at this time can be staggering, such as the 20,000 Redwings which were counted on a single October morning passing over the highest point above a town near me. You certainly don't need to head to coastal hot-spots to see migration at its best in autumn.

Visible migration involves seeing, and hearing, birds that departed points either on the continent, or further away in the UK. It is very easy to overlook, but once you're aware of it, you will be

Diary Notes:

15th September – Top of the hill

Laura and I went blackberrying around the lagoons at the reservoir today. The hedgerows are a reliable spot for a good crop of juicy berries and it is a handy circular route of a mile or so. It was no use pretending that I didn't know two juvenile Spoonbills had arrived there, though, so I fessed up to Laura regarding my urgency to do it today. After filling our baskets, we both enjoyed the sight of two of these unusually active individuals sweeping their spectacular spatula bills through the shallow waters of one of the pools at point blank range. We came home with a good few bags of blackberries too, so everyone was happy.

HOW TO:
SEE BIRDS MIGRATING

Millions of birds, including Woodpigeons, Starlings, finches and buntings pour into the UK between September and November. Seeing them arrive is a thrilling experience.

• The best 'visible migration' watching comes at the coast, where birds follow the coastline before turning inland, and on hills, so head for the highest point in your patch.

• The direction in which migrants fly through your patch can vary. Once you have worked out if you are on a north-south flyway (as I am) or an east-west flyway, you'll know which way to look.

• Birds fly lower into a headwind, so if they are heading south in autumn, a southerly may keep them lower.

• Brush up on your bird calls as many birds will be silhouettes, so you'll need to use flight calls for ID.

• The first couple of hours after dawn is the busiest period, so set your alarm.

Great Skuas are on the move in autumn and gales can bring them inland.

amazed at not just how many birds you can see heading purposefully overhead, but also the variety of birds on the move, from Meadow Pipits and Chaffinches to Fieldfares and Lapwings. You might think that birds such as Starlings, Woodpigeons and Skylarks are residents that rarely move far. However, these three species pour into the UK on many days in late autumn. They are easy to miss, unless you are aware of the size of the populations of resident birds of each species in your local area.

Similarly, if you know the regular roost flights of birds such as Woodpigeons and Starlings around your patch, you can be sure if immigrants are arriving – from as far away as Russia in the case of the latter.

You won't witness Goldcrests visibly migrating, unless you live by the coast and see them coming in off the sea, but a sudden increase in numbers in your local wood or park in October, or finding them in spots you wouldn't usually, means there has been an arrival from Scandinavia. Many will stay in the UK throughout the winter.

Take early morning walks on playing fields and larger lawns and you will find that Blackbirds suddenly become more numerous on some autumn mornings. You'll also notice very shy Song Thrushes zipping overhead and plunging into cover, or exploding with sharp 'tsip' calls from ditches. The reason is that immigrants have arrived from Scandinavia.

Spring migration has the edge when it comes to colours and fine plumage, as the birds are adults in their breeding plumage, but autumn has the year's flush of juveniles, so there tend to be more birds around. Juveniles are more prone to wandering than adults, meaning that you could encounter something very special indeed, perhaps even a bird from the other side of the world. Birds that have bred further north in the UK pass south through the country, but you'll also get birds that have bred in northern Europe and drifted off course, perhaps even an American or Siberian visitor such as a Pectoral Sandpiper.

Juvenile birds can be incredibly confiding. In the case of some Arctic breeding species, including many waders, you might well be the first human they have ever seen in their life! This means that they have no fear whatsoever and you can get some fantastic views.

A huge variety of migrant birds will be

Weighing no more than a ten-pence piece, migrant Goldcrests will arrive on your patch in September and October.

passing through your local area and knowing the best spots to look will help you to connect with them. Certain birds use the same spots every year, and no doubt pass

on this information to their young ones.

Birds that have bred further north such as Whinchats frequent open ground with scattered cover and vegetation. Barbed wire fences are perfect perching places for this insect-eater, so keep checking spots where you have seen birds previously. The advantage of fences is that they are usually there the following year, so you can track down regular spots.

Keep an eye on flocks of small birds in autumn – gems are lurking. Every autumn Yellow-browed Warblers are turning up in greater numbers from their Siberian breeding grounds, so keep your ears open for their surprisingly loud and far carrying 'tsoee-eet' call. Some mixed flocks can be truly spectacular. Tit flocks can number more than 100 birds and take an age to pass through a hedgerow or a single big oak tree. They often have Treecreepers tagging along. See how many different species you can record in one flock.

Late August to October is a time when you have a great chance of being visited by something unexpected. Autumn's mass migration could bring something a little bit special, so stay alert. An incredible number and variety of scarce and unusual visitors turn up in gardens every year. A Wryneck feeding on black ants on our cracked patio, a Black Redstart on the garage roof, or a Pied Flycatcher in

The stunning Firecrest, the Goldcrest's rarer cousin, could come your way in autumn.

Your local wetlands may tempt a Whimbrel to drop in.

the apple tree are birds I hope for each autumn

Migrants dominate in autumn, but during this season one resident that will become much more obvious in your local area than at any other time of year is the Jay. They play an essential part in oak regeneration. They are busy collecting acorns and stashing them for eating in hard winter weather, so make regular long flights across open areas. If the acorn crop is poor in your area, they may well move away completely.

NOCTURNAL MOVEMENTS

A lot of migration also goes on under the cover of darkness, but that doesn't mean you can't witness it. Unfortunately for me, the main road is a little bit too close to the house so I don't have the best spot for nocturnal listening. That said, I have heard flocks of Wigeon arriving under cover of darkness, while in the bath, and once, bizarrely, a Water Rail 'singing' as it flew around in the dark.

Try the same tactic for waders. It's well worth just sticking your head out

The starling-sized Grey Phalarope could arrive on your local lake or reservoir after a big blow between September and November.

of a window for a few minutes, or heading outside into the garden for a quick listen. You might pick up the distinctive 'seven whistles' of the Whimbrel, the ringing 'pee-tweet-tweet' of a Common Sandpiper, or the 'chu-chu-chu' of a Greenshank.

The classic incoming bird from mid-September through to November, with one of autumn's most evocative sounds, is the Redwing. Keep an eye on the weather to see if a clear night is forecast because that's when you should get outside and listen for the thin lisping call. Flocks of Redwings use the stars to help them navigate.

For many, the first 'seeh' of a Redwing is the first sign of winter. It's a sound that

Look for returning Green Sandpipers from July.

is perhaps not as eagerly awaited as the first Swallow of spring, but the changing of the seasons is something we can't control so welcome it in.

WADER WONDERLAND

Waders are among the longest distance migrants of all and their autumn journeys can involve thousands of miles of travelling. Track down suitable wader habitat in your patch — insect rich, oozy mud and shallow water entice tired and hungry waders to make a pit stop on their way to Africa between July and October. Water levels at reservoirs and lakes often drop in hot summers, creating ideal feeding grounds for a variety of both long- and short-billed waders which are hungry from their long migration. In contrast to the spring, when the race is on to get to breeding grounds first, there is no hurry to move through so you can enjoy the birds for longer.

The timing of when you see birds can give you a clue as to whether you are watching an adult or a juvenile. You'll find that the first arrivals for each species tend to be adults, still in slightly worn breeding finery. The earliest are often failed breeders. The crisply marked juveniles, decked out in a spanking new set of feathers, are a real treat. Look closely and you'll be dazzled by an array of chevrons, bars and tiger-stripes. The first juveniles occur at roughly the same time each year, some time after the adults – for Siberian waders, including Little Stint and Curlew Sandpiper, it is usually the second half of August before they turn up.

Autumn wader flocks are great for harbouring vagrants, so check Golden Plovers in the fields and Dunlins on the marsh for something special, perhaps an American Golden Plover or Buff-breasted Sandpiper (I've seen one of each on my patch).

HOW TO: FIND HOBBIES

Hobbies are one of the last birds to finish breeding each summer. Mid-August to mid-September is the time to get out for the best Hobby viewing of the year.

• Listen. They are incredibly loud and vocal and calls provide a real helping hand in tracking them down. I once found five nests in as many days on my patch thanks to the persistent calls of noisy juveniles.

• Hobbies will often use the same nest each year, frequently in an old crows' nest, so revisit the same location the following year. The chances are they'll be back.

• Juvenile Hobbies spend long periods sitting in trees and waiting for the parents to bring them food, such as Migrant Hawker dragonflies – the peak flight time for these insects coincides with the Hobbies' fledging period.

MAKING HOMES FOR WILDLIFE

Bird activity, and numbers in your garden, will be less between August and October, but there's no need to worry. Seed-eaters, such as sparrows and finches, will be out in the fields and hedgerows taking advantage of the natural bounty, so don't fret if numbers of birds at your feeders are down. Just scale down the feeding.

On the subject of seeds, it's the time

Hobbies nest late, so can feed their young on Migrant Hawker dragonflies in autumn.

to gather seeds from plants you'd like to grow in the garden and to make friends with your neighbours and maybe ask them if they'd mind you taking some seeds from the plants you've been admiring all summer.

Leave Sunflowers, Teasels and other good seed-bearing plants where they are, so they can provide seed for winter finches over the winter once they've returned from the fields. Some birds will still be around, though. If it has been a good summer, House Martins may still be busy with finding food for their young well into September. They don't nest on my house yet, but if you have them, it's great to watch them taking in food and the young peeking out over the edge of the mud-cup nest.

In the garden it's a time for a bit of a tidy up, but a lot of what you do has an impact on what happens next year. Leave longer stems of woody plants in the garden when pruning in autumn. The hollow interiors can house dozens of creatures hibernating over the winter, including ladybirds and

Diary Notes: 15 October – The big blow

The gale-force northerly winds overnight resulted in some exciting birding with a lot of seabirds being blown inland. A dawn visit to the reservoir was a must to see what this magnet for migrant birds had pulled in, and in an air-punching moment of satisfaction, a grumpy looking Great Skua greeted me. It sat on the water in the middle of the reservoir. As it turned out, the Bonxie was just the appetiser. I found myself rushing home after work for seconds to see what else had arrived as more and more reports of inland seabirds rolled in. As the Black-headed Gulls assembled this evening in the usual roost, a 'black one' was suddenly sitting among the carpet of white, causing a rapid quickening of heartbeat. Its delicate build and cold plumage tones revealed it to be my first Long-tailed Skua in the county – a prize find anywhere, but especially so a couple of miles from home in landlocked Cambridgeshire.

The following morning, as others failed to see the skua, which departed at first light as the gulls moved off, I found a tiny Grey Phalarope bobbing around on the deep water, spinning in circles as it fed on insects off the north shore. What a couple of days and all thanks to the storm.

House Martins may still be raising a brood well into September or even October before departing for Africa.

QUICK TIP: MAKE A HOME FOR HEDGEHOGS

Hedgehogs are sadly in trouble with a serious decline in the UK in recent times, but you can do your bit to help them. Leave gaps under new fences, or make a hole in existing ones so that they can travel between gardens – and into yours. Make piles of branches, vegetation and leaves in quiet spots in your garden so that they can hibernate. Never use slug pellets in your garden.

Common Earwigs. Autumn leaves can be a pain as they accumulate in all the nooks and crannies you don't want them in, but cram some into log piles and other spots in the garden because they're good for hibernating creatures too.

It's the perfect time for picking up unwanted sections of tree trunks, logs and branches. If you see any neighbours or local authorities doing any tree surgery, keep an eye for any leftovers or better still, ask if you can have some of the off-cuts for your garden. Before you know it, you'll have built up enough over

Drill holes in old logs for solitary bees and wasps.

time for a magnificent log pile! The more piles you have the better – we have half a dozen around the garden.

Logpiles look great in the garden and will seriously boost the potential of your own plot to host wintering reptiles, mammals and amphibians, plus hundreds of other creatures. Drill a good number of holes in the ends using different sized drillbits and you'll also attract bees and solitary wasps going in and out in summer. Deciduous trees are best, but you can mix in a few coniferous sections for variety. Once you have established piles, have a look occasionally to see what's living there. Several species of woodlice should move in, plus earwigs and centipedes. Fungi will colonise, as will moss, and you'll soon have a very natural woodland feature in your garden.

Old pallets are also very useful when it comes to making a wildlife garden. Stack several on top of one another and fill the gaps with flower pots, old bricks, logs, sticks, leaves and anything else you can think of that might make a good home for wildlife. An insect hotel will be a real centrepiece for your garden and is great fun to make as well. I managed to obtain several pallets that were being thrown out at my gym and at work and was told to take them away with the owner's blessing.

Putting nestboxes in your garden is very valuable for wildlife as natural nesting and roosting sites become fewer. We gladly 'lose' one of our bird boxes on the back of the house each year to Tree Bumblebees.

HOW TO: SEE HUMMINGBIRD HAWKMOTHS

This gem of the UK's moth fauna has an incredible turn of speed. I've found two sorts of flowers very attractive to 'HBHMs' in summer and early autumn – Red Valerian and the richly sweet-scented purple or white flowers of Buddleia davidii. You should certainly grow these in the garden, but when summer arrives have a look further afield to see where they grow and go in search of this quite brilliant little migrant moth.

Create a variety of nooks and crannies in your insect hotel.

QUICK TIP: FINDING NESTS

As the leaves drop off the trees and hedgerows in autumn, look out for bird's nests. Leave it too late in the year and they'll start to decompose. One nest that is well worth seeking out in bushes is the ball of the Long-tailed Tit. They are made of lichens, cobwebs and feathers and the cobwebs allow them to stretch as the young grow – this is a magnificent feat of engineering. It's also a good time of year to count the nests in your local rookeries and heronries.

When they arrived with us, they were still a relative newcomer to the UK. They are great to watch as they zoom in and out of the box and we sit having dinner outside as they go about their business.

INSECTS – IT'S NOT OVER YET!

You can still enjoy insects well into autumn before the first frost does its deeds. Autumn is in fact a great season for moths, with some spectacular species taking their turn on the wing from the green-and-black Merveille du Jour to Frosted Orange and Centre-barred Sallow. Seeing these three alone should dispel the myth that moths are brown and boring. Moths also have some of the best names – for example, the Setaceous Hebrew Character, Mouse Moth, Maiden's Blush and The Snout – and I think such memorable names help you to remember them.

A simple sound that I associate most with autumn's shortening evenings is the single chirp of the Dark Bush-cricket coming from deep within hedgerows and bushes. Almost everyone will have heard it and hear it every year, but I bet not many know the source. Try and track them down and get a view of a common, but never-easy-to-see, insect. They can go on well into October and are a key part of the autumn evening atmosphere. Trying to hear that chirp a day or so later than the previous year is another challenge.

You don't need to have a pond to be

This nestbox in my garden is popular with Blue and Great Tits.

HOW TO: USE NESTBOXES

Put nestboxes up in autumn because birds will find them and start investigating them at this time.

• Position your bird boxes facing east. North is too cold, south too hot and west exposed to wind and rain.

• Hole-fronted boxes will attract Great and Blue Tits; open-fronted are good for Robins, Wrens and maybe Spotted Flycatchers.

• Try an old tea chest in a mature tree for Tawny Owls.

• Put up some bat boxes too.

• Hole-fronted boxes, or even a big flowerpot on its side, filled with hollow bamboo canes, will attract bees.

Blue Tits will provide hours of entertainment if they adopt a nestbox in your garden.

able to enjoy dragonflies in your garden in autumn. Migrant Hawkers are one of the last species of the year on the wing and they spend a lot of time patrolling the skies. You can expect your first sighting of the year by late July and they can then be found regularly well into October (the latest date I've seen one is 16th November). This is one of the few species of dragonflies that patrols in flocks. Find

HOW TO: SEE RAPTORS AND OWLS

• Listen for the sudden trill of a Blue Tit piercing the air. This often leads you to a sighting of a Sparrowhawk coasting menacingly overhead.

• Scattering Woodpigeons and ducks are the clue that a Peregrine is on the hunt.

• Alarm-calling Swallows and scattering House Martins are a sign that a Hobby is around.

• Flocks of finches, larks and pipits suddenly bursting from the ground into the air reveal a Merlin, or perhaps even a harrier quartering the area.

• Listen out for persistent 'chinking' from Blackbirds and 'pinking' of Chaffinches. If they continually fly into one spot, the chances are that there is a day-roosting owl. Tawny Owl is the most likely, but you could find other species too.

Dark Bush-crickets 'chirp' on autumn evenings.

Migrant Hawkers are one of the last dragonflies on the wing each year.

Hummingbird Hawkmoths hover at garden flowers just like their bird namesakes.

a sheltered spot in your local wood, or even over the garden, and you could see up to 20 of them feeding together, jinking and jerking away from one another as they claim their own airspace.

A washing line might not spring to mind as must-have habitat to provide, but they are great perching places for Common Darters – another dragonfly species that occurs well away from water.

MINI MIGRANTS

Migration isn't just restricted to birds. Many moths and butterflies are long-distance migrants. 2009 provided one of the most phenomenal experiences for nature lovers in the UK in my lifetime. Painted Ladies arrived in their millions from North Africa that year and gardens, wildflower meadows, in fact anywhere and everywhere, were full of these beautiful butterflies. I spent half an hour counting Painted Ladies migrating rapidly over a high-elevation car park near me and then scaled up the number passing through that 100-metre stretch to cover the whole area. This gave a figure of around half a million butterflies potentially moving through my patch.

Many national events, such as that well-documented invasion, are apparent at a local level, so keep an eye on what's happening around the UK through reports on the internet or in the national press. Invasions of Clouded Yellows, the most brilliant yellow of any wild creature in the UK, are often reported in this way. They are fast flyers and rarely linger for long in one place. I once found one in early November, so this is a butterfly that's well worth looking for long after

Painted Lady – a common migrant one year, rare the next.

the others have finished flying. In some years, fields, grassland and meadows in southern England can be full of them so keep an eye open for news of their arrival.

The species that I'd probably most like to discover on my patch is the Camberwell Beauty. I love the idea that rarities can turn up in even the smallest garden and this spectacular butterfly is one of them. Easterly winds in autumn mean that they can arrive from Scandinavia and the butterfly bush, the Buddleia, is as good a spot as any to discover one. Keep looking!

Another rare migrant butterfly that has started to turn up occasionally in early autumn is the Yellow-legged Tortoiseshell, so make sure you are familiar with the widespread Small Tortoiseshell.

Head outside with a torch in August

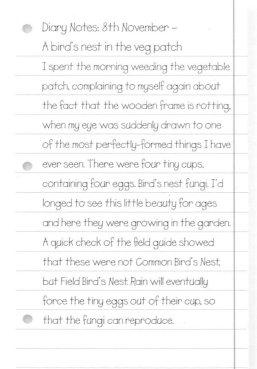

Diary Notes: 8th November –
A bird's nest in the veg patch

I spent the morning weeding the vegetable patch, complaining to myself again about the fact that the wooden frame is rotting, when my eye was suddenly drawn to one of the most perfectly-formed things I have ever seen. There were four tiny cups, containing four eggs. Bird's nest fungi. I'd longed to see this little beauty for ages and here they were growing in the garden. A quick check of the field guide showed that these were not *Common Bird's Nest*, but *Field Bird's Nest*. Rain will eventually force the tiny eggs out of their cup, so that the fungi can reproduce.

Look carefully on old wood for tiny bird's nest fungi.

to check out your buddleias for another immigrant: the Silver-Y moth. In good years, you could find a dozen or more nectaring at the sweet-scented flowers.

Rare dragonflies are a feature of early autumn. Lesser Emperor, Red-veined Darter, and in some years Yellow-winged Darter, can arrive, so check all your Emperor Dragonflies and darters carefully. A prolonged spell of warm south or south-easterly winds may drift them here.

Dewy autumn mornings are a time to enjoy glistening spider's webs. Some common species to look for are the Four-Spotted Orb Weaver and the Garden (or Cross) Spider. There are somewhere in the region of 650 spider species in the UK. Some other easy ones to look for during the year are the small, black-and-white Zebra Jumping Spider, wolf spiders and crab spiders. Stilt-legged harvestmen and craneflies are another feature of autumn mornings.

Sunny autumn days are a good time

wood and leaf piles for them, and other creatures, to hibernate in.

As Christmas comes, it's the time to reflect on your highlights, the things you saw and the things you didn't; the jobs you did in the garden and the ones you need to plan in for next year. One thing is for sure, no two years are the same on anyone's patch. Who knows what next year will bring?

Common Garden Spider, or Cross Spider – even the commonest species are worth a closer look.

to check your local fruit trees and watch insects getting drunk on the fermenting fruit. Get to grips with the several species of wasps, including the biggest and most spectacular of all, the Hornet. This is a gentle giant, so don't be alarmed by its size. Its markings are stunning.

As the weeks of autumn progress and the days shorten, mammals step up their feeding and fatten up for the winter ahead by gorging themselves on autumn's bounty of seeds, nuts and fruit. Hedgehogs are really active at this time and you can do your bit by collecting leaves and making

Autumn woodland's riot of colour is a highlight of the year.

A Kingfisher sighting will brighten any day.

Diary Notes:

31st December – A lunchtime treat

We walked from home to the mill and back today with friends. No Otter, as usual, but a Kingfisher performed brilliantly which was great for one of our group as she'd only seen one briefly before. Not only did it give the classic flypast, it even perched on a pollarded willow stump and fished right before our eyes, emerging with a tiddler. The sun was out and it was probably this that tempted a Lesser Spotted Woodpecker to give five bursts of its falcon-like trill from a small poplar plantation on the other side of the river. A new local site for this increasingly scarce species. Things were to get even better, though. We were enjoying lunch and a pint when I saw a crest appear in a cheese plant right outside the window. I was about to launch into my usual spiel of how the Goldcrest is our smallest bird when it popped up in view to display a black-and-white eye-stripe – a Firecrest.

It then hovered at the window and feasted on spiders as we sat there eating our lunch just four feet away. This bird hung around for several more days for other local birders to see. Proof that the most unexpected things often turn up in the most unexpected places.

ACKNOWLEDGEMENTS

I owe a great deal of thanks to the many people who have helped me to discover and learn about the amazing wildlife to be found in the UK. There are far too many to name here but for everyone who has answered my questions, or shared information that has allowed me to see something special, thank you.

To my good friends, Ade Cooper, Richard Patient and Alan Hitchings, with whom I have shared many an adventure and some excellent birding in my patch and further afield. To Dr Mark Gurney – one of the best, most knowledgeable and most modest, naturalists I know, with whom I have shared great days tracking down the UK's rarest flora. To Bedfordshire Hoverfly Recorder John O'Sullivan who tutored me in hoverflies, and to my local fungi groups who have allowed me to join them and learn about this fascinating family from their experienced members.

I'd like to thank my parents for carefully nurturing my interest in wildlife from a young age and being so encouraging with a hobby that has also led to a career in natural history. It is a lucky person indeed whose passion is also their job.

Finally, I must thank my wife Laura who has shared in many of my adventures in our first few years in our home and will share in many more in the years to come. I'd like to take this opportunity to apologise for deciding that sorting out the garden was more important than cleaning the house and tiling the kitchen when we moved in ...

IMAGE CREDITS

INDEX